42—Th

"Behold! Human beings living in a underground den… Like ourselves… they see only their own shadows, or the shadows of one another, which the fire throws on the opposite wall of the cave."

Plato 429-347 BC

www.2x21.com

ONVIEW BOOKS
Published by Onview.net Ltd.
2007

Registered Office:
Onview.net Ltd. 4a Oxford Road, Wallington, Surrey. SM6 8SJ. England.
Distribution:
Frilford Mead, Kingston Road, Frilford. Abingdon. Oxfordshire. OX13 5NX. England.

www.2x21.com

First Published 2007 by Onview.net Ltd.

A CIP catalogue record for this book
is available from the British Library

ISBN 978-0-9557137-0-5

42

The Answer to Life,

the Universe, and Everything.

by Mol.

This book is dedicated to the 6 unknown ants I incinerated with my sun-magnifying glass heat ray in the summer of 1962 at 2 Links Rd., Tooting, when I was eleven years old, and to the future intellects of machine-human hybrids and their evolving consciousness.

Sorry ants. Because you were different to me, I thought you were unimportant, but I know better now.

Hello machine-intellect and welcome to the issues. Remember the ants!

Acknowledgements & Credits

Writing a book about science, religion, and the kind of material sensitive to most of the 6.5 billion people on the planet is a difficult goal. I am obviously aware of the strength that spiritual ideas and scientific theory have in their fundamental hold on our sense of purpose and understanding reality. I have tried to retain the courage of my convictions, and say it how I see it to be with no holds barred, in an attempt to wrench people away from what I consider to be pre-programmed and out-of-date ideas; many of which, have not served humanity very well. This work is bound to upset many readers, but this is not my aim. My intention is to encourage everyone to look at their own internal hard-wiring and to challenge how it was made before they had any opportunity to look at life with fresh and unfilled minds as children. We are not born into a world of new ideas: we are born into a swirling torrent of other people's ideas—dead people, and we are systematically programmed to become living beings carrying the thoughts of past generations. Many of *their* thoughts were wrong, and only a few remain valid today. *We* are not given any opportunity as infants to look for truth concerning our most powerful questions. Why? Because we are invariably answered by well meaning parents intending to provide us with their moral guidance. Alas, such well-meaning actions contaminate our capability for determining reality with stories only appropriate for generations of two thousand years ago. I believe we should not be taught to have unwavering faith in anything except the idea that we have fresh opportunities to look again, at what is true through the process of being born idea-free!

I asked many colleagues and friends to review this work before releasing it into the big wide world. Each of the reviewers came from different backgrounds, and held varying beliefs to one another. I would like to thank all of them for their positive critique and feedback. Their help has enabled me to remain, more or less, on a subject dear to us all. Through the guidance of good, caring, normal people, my thoughts now have an opportunity to be considered by a wider audience. I have tried hard to be informal in my presentation of difficult concepts as a result of their genuine interest in wanting to consider the reality we all confront and think about. Thank you for listening.

I especially thank:
Lesley Evans, for her love and encouragement as much as for her very long hours checking my work and removing typos and errors of grammar; Michelle Williams for being the first to review my work, providing excellent critique, and giving me confidence in what I had to say.

Mol

Preface

Who am I? Where do I come from? Is there a God? What happens when I die? How should I live my life? Is there a plan? Does fate exist? What is my purpose? Did I arrive on the wrong planet? Who is this book for? Is it for me?

Will this book really answer my questions?

YES!

Science versus Religion: ***The Don Delusion?***
Scientific reason is often straight-jacketed by institutionalised indoctrination, and refuses to become infected with the ideas, and perceived absurdities, of religion. Theologians, equally afraid of their position and the values they wish to protect, launch repeated academic defences against the mutually exclusive attack from science. Discerning people throughout the world, eager to learn the nature of all things and their part to play, are torn to this side or that: God, spirituality, religion, on the one side—science, knowledge, and heartless universe on the other. Are our lives only the product of ruthless laws in a physical universe, devoid of abstract intent? Is the answer for human purpose only to be found in the out-of-date ceremonial ritual of a two thousand year old religious ideology? Can science, reason, and new ideas, not redefine a clearer insight into our nature of existence? Is there nothing new to answer our probing intellectual questions and our innate feeling of deeper meaning for our lives?

Yes. There is.

A few eminent, educated, and qualified people, who should be casting off their blinkers to talk with each other to combine their convictions and work, are not doing that. They are fighting over their differences. Their bruised egos frequently imbue their writing not with considered, and objective, debate but with scorn and ridicule aimed at character assassination of their opponent. Since the learned professors and dons, with their combined knowledge of hard science and spiritual beliefs, are locked in a spitting war, and thus unable to offer a convincing, unified, account of what everything is all about, I thought it is about time someone tried. But then are we only to explore the nature of everything through the polarised perspectives of two camps, or should we not use broader skills: art, creativity, and intuition too?

I think if we are to discover the meaning of life, we require more than just a few narrow views to comprehend it. I have elected myself for the job of trying to illuminate the truth. What I have to say already exists in various books written by good authors in different fields of science, art, and philosophy. I have merely tried to pull all the parts together to present a reasonable set of solutions to the unanswered questions. If I have added anything of value then it is only open-mindedness, and the refusal to accept anything I am told to be true unless it makes sense in an otherwise absurd world.

Mol

*   *   *   *

Contents

PREFACE 5

SCIENCE VERSUS RELIGION: *THE DON DELUSION?* 5

CONTENTS 7

INTRODUCTION 10

CHAPTER 1: SEEING CLEARLY 18

ME 18

US 21

THE POWER OF DECEPTION AND FALSE BELIEFS 23

A POLICE FORCE WAS CREATED TO PROTECT THE POPULATION OF A CITY OR COUNTRY 25

CCTV CAMERAS PROTECT YOU 26

ANARCHY MEANS REVOLUTION AND CHAOS 26

GLOBAL WARMING IS MAN-MADE 26

CHAPTER 2: ANT MINDS 28

CHAPTER 3: SUPER MINDS 36

CHAPTER 4: BIRTH 42

CHAPTER 5: THE ILLUSION OF SELF 47

CHAPTER 6. TO BE OR NOT TO BE... MORE, LESS, OR NOTHING 58

CHAPTER 7. DEATH OF HUMANITY 64

CHAPTER 8. TAKING THE CHRIST OUT OF CHRISTIANITY **69**

CHAPTER 9. CHILDREN OF STARS **80**

THE LINE OF CONTINUITY – BIRTH AND REBIRTH 80

CHAPTER 10: DESIGN **86**

INTELLIGENT DESIGN? 86
ANOTHER SMALL TALE: 88

CHAPTER 11: THERE BE MAGIC TOO! **93**

ENTER THE NOTION OF DARK ENERGY! 94
MORE MAGIC: DARK MATTER! 94
QUANTUM THEORY 95
MANY WORLDS THEORY/MULTIVERSE 97
WAS GOD AN ASTRONAUT? 100
DID THE ONE TRUE GOD MAKE US? 100
ARE WE REALLY LIVING INSIDE A COMPUTER PROGRAM? 102
WAS OUR UNIVERSE JUST SIMPLY MADE IN A LABORATORY? 102

CHAPTER 12: LOCAL LIFE **105**

LOCALITY AND YOUR PART IN THE NATURE OF THINGS 105
CHILDREN 105
ADULTHOOD 107

CHAPTER 13: THE MATERIAL WORLD **113**

WHAT IS WEALTH? 113

CHAPTER 14: DEATH **120**

WHEN YOU DIE, YOU ARE DEAD 121
HEAVEN AND HELL 123
REINCARNATION 125
WE ARE THE DREAM OF A HIGHER INTELLECT 126
WE ARE THE DREAM OF OUR FUTURE SELVES 126
ALIEN LIFE FORMS MADE US 127
WE ARE IN A COMPUTER PROGRAM AND THEREFORE NEVER DIE 127

CHAPTER 15: PREPARING FOR THE ANSWER **132**

CHAPTER 16: THE DECISION **141**

THE END OF DAYS 141
THE RISE OF THE MACHINES 144
THE END OF EONS 147
RAISING THE DEAD 148

CHAPTER 17: THE ANSWER TO EVERYTHING—42 **152**

THE RECKONING 152
TO BE OR NOT TO BE 153

CHAPTER 18: EPILOGUE **155**

A FINAL THOUGHT TO PONDER 159

APPENDICES **160**

I FALSE BELIEFS (FULL LIST). 160
II OBSERVATIONS 168

REFERENCES, FURTHER READING, NOTES **169**

*1 BIG BANG UNIVERSE. 169
*2 BRANES. 170
*3 OMEGA POINT 170
FURTHER READING 172
ABOUT THE AUTHOR 177

Introduction

Everyone eventually asks the question, "Why am I here?"
Many people have already discovered a wealth of material aimed at answering their most profound query. Unfortunately, religious theories, blind-faith belief systems, touchy-feely quasi-philosophical ranting, along with distorted scientific truths, unimaginative logical dogma, and fashionable hype fill the vacuum in the human-mind—not with truth, but confusion, conflict, and illusion. It is no surprise to discover the shelves of retail bookshops are filled with a proliferation of titles under the heading of 'Spiritual'. Closer inspection reveal books on every aspect of self-help and how to enrich our experiences. It appears many members of the human-race have reached a point where they desperately seek genuine purpose rather than remain, for the most part, unfulfilled. This phenomenon exists predominantly in western society, where giant super-markets fill cathedral corridors with the greatest hoard of material wealth ever assembled in one place; where each minute of every day, citizens are bombarded by bill-boards, television screens, emblazoned buses, banners, newspapers, magazines, and store windows to own this thing or that to improve their routine existence!

Outside the western world, two-thirds of the six billion people on the planet struggle to obtain their daily needs. Most are on the bread line wondering where the next meal is coming from. Their spiritual questions are answered by inappropriate religious indoctrination, designed only to convince them to come to terms with their misery in this life through the false promise of a happier existence in the next one, or by western propaganda infiltrating their societies and bringing, to them, the same deceptive material-promise we already know.

Apart from illness and human-intervention, most other animal species on the planet do not seem to experience the kind of unhappiness we do. A tiger appears to be fundamentally content with its lot. Predator and prey live in better harmony than people do. One could argue this is because we are thinking bipeds and are 'aware', whereas other animals are not. We are able to think intelligently. We have the ability to deceive and lie. Such argument fails miserably in the light of the fact many animals, among them octopus, dolphin, monkey, elephant, dog, fox, and wolf, have proven intellects and mental capabilities equal to our very young infants. It is difficult to prove what 'self-awareness' and consciousness actually involve, but I believe a tiger is happy with his lot because he knows, inherently, he *is* a tiger. He behaves one hundred per cent to his role and purpose. All other creatures on the planet sufficiently aware to explore their environment, intellectually, know what they are, and have little need to question it. A fish enjoys being a fish, because it will never be anything else, and therefore does not worry about it. It *knows* what it is either because it lacks the imagination and capability to wonder, or because it is born with inherited traits to be only a fish.

People are unhappy for different reasons, but I believe that underlying all of them is a single cause: they have never discovered or understood their unique role in being alive, and do not understand what their function is. When we ask our academic peers or our moral leaders for intelligent advice, we are directed towards ancient soothsayers, unproven deities, gods, prophets, dusty-texts, and muddled books of wisdom for our answers. It's like asking grown-ups to believe in tooth fairies. People are given stories only suitable for the inexperienced minds of young children or, just as bad, unfathomable theories suitable only for the membership of MENSA. Faced with myth and fable in answer to our questions, we remain in a state of ignorance about our role in the grand evolution of a universe. We are left to wander through short lives grasping at being all things, but never truly being the one, we were designed to be. It is as though we are tigers, believing we are elephants, with all the pain and disappointment it brings in not being able to discover our true identities.

Human history is populated with emerging ideas regarding religious belief. Paganism, Hinduism, Buddhism, Islam, and Christianity are just a handful of repeated attempts at determining a shared universal purpose. Did we evolve or were we created by a supreme intellect? Are our lives just proving grounds, testing our loyalties to this God or that one? More recently in the last four hundred years, an ancient ideology achieved sufficiently new prominence to gain worthy attention again: *science!* From its outset, this strand of human understanding has required repeatable, demonstrable evidence to be forthcoming for its arguments to be accepted as true. Religion, on the other hand, has never required repeatable proofs; instead, it asks only we believe, and many of us do exactly that—believe! We do so in defiance of the cold and ruthless light of reason. Why?

Faith and religion are often cited as having profound effects on people's lives, with many believers reporting dramatic life-changing moments as though these experiences were somehow fated to happen. Bad people have become good people. Good people have become amazing benefactors to other people, whose own lives are blighted with poverty and disease. There is no doubt that passion can drive human endeavour, and enhance the idea of being on a righteous path serving God. Yet, such behaviour offers no proof of any religion being true; people can perform similar acts, and feel overwhelming emotions, through passion and belief in their country, their team, or their own ideas. Believing in something, no matter how great it makes you feel, or how much it seemingly provides strength and purpose in a harsh and difficult world, does not offer evidence for any spiritual influences—it just illustrates the way emotional traits, within our makeup, can help us feel better, or drive us to make positive contributions in our lives. If we possessed a different mix of chemical stimuli from the norm (one questions what is normal), our feelings would be different to those we currently have. Our minds, emotions, and our body chemistry are woven together into an inseparable mix, which provide each of us with a *subjective* view of life. There is no objective view of anything. Humans seem unable to untangle themselves from their own, poorly constructed, version of universal truth. This is because we are all effectively deluded in one way or another by other people even more deluded than us.

A spate of books has been published setting faith-believer against pragmatist—religion against science. What appears lacking to me in most of these informed, and cleverly argued, works is recognition of the boundaries separating the diverse strands of human need. We seek knowledge but also purpose; we comprehend mortality but still harbour hope. We are able to imagine fantastic and impractical scenarios on what we really expect to happen when we die, simply because we are unable to accept the finality of death. We secretly know much of what we dream and hope for can never possibly be.

People live for more than material security and the daily practicalities of getting by. Art, music, love, musing, daydreaming, fooling around, and humour are equally important to us as knowledge, wisdom, and morality. The theologian amplifies the importance of spirituality in fulfilling human abstract need, whereas the scientist offers a growing understanding of reality, but one devoid of any promise to fulfil our fundamental yearning for a greater purpose.

Religion and science, instead of being opposing forces may, in truth, be parallel ones failing to recognise their own origins and the different methods they employ to aim at the same target. The problem with specialisation is that, when one becomes immersed in a thought discipline, most other ways of seeing a thing become obscured. Where scientific knowledge of the universe, and our place within it, is weakest and least proven, the champions of religion fill the void with mystique. Scientists fall victim to the same mistake as the theologians, and pick through the detail of biblical text trying to find fault, instead of considering how their own area of study might offer fantastic spiritual possibilities for human purpose. Both camps resort to exposing the gaps and weaknesses in their adversaries' arguments, when they should be accepting that we all have a long way to go, before we obtain absolute understanding of everything.

Science is the slow deliberation of acquiring knowledge and truth. We are only a small way along all the learning paths, and we still have so much yet to find out. Science may, one day, ultimately discover a divine awareness (God or gods) responsible for all creation, despite the fact there is no hard demonstrable evidence of this right now. Religion appears to satisfy the need for purpose—human purpose, something casually reduced by science to be a 'purpose-less' role: we are just a chemical reaction in the universe and have no pre-determined point in being. The problem is religion offers a promise of purpose within an absurd modern framework, and from the shaky foundation of a chequered history; it is devoid of modern thinking, and attempts to sell an old idea beneath an atmosphere of well-deserved cynicism from science-enlightened people. Science offers only a cold and convincing knowledge of life, but adds nothing of value to help us accept the struggle, and the terror of death. Are these really the only two philosophies humankind has to make choices from after 4.5 million years of existence?

The questions we should be asking are not whether religion or science holds definitive answers to our human curiosity and emotional needs, but whether these divided camps can offer a convincing method for finding out. I think the best way to attempt to create a proper resolution to the meaning of everything, and the discover the purpose of life, is to explore how scientific truth can be wedded to the core values of spiritual ideologies, irrespective of whether the answers are acceptable to purist theologians, or unimaginative scientists. Hopefully, a large percentage of open-minded ambassadors for both of these disciplines will glimpse the seed of unification and glaring truth in this account.

I believe we have always been lost. Humanity gropes in the dark. After a few thousand years of stumbling around, a few small truths have been discovered to shed sufficient illumination on a unique and almost impossible existence. I think enough data has been gathered for each of us to find the real light switch, and the time has come to resolutely flick it on! This book will explore and expose exactly who we are and suggest what our true purpose is. It is not about belief at this stage; it is simply carefully deduced argument. We live in deceptive times where it has become far more difficult than in the past to prise out the truth from what we are told. I want to blow away the dust of religious fable and remove the vain mirror of human distortion from science. I want to tease out the facts from what is known, and present answers to the meaning of life clearly for everyone to understand and consider.

My book would be incomplete if I dealt only with what can be proved today. Knowledge is continuously being added to our understanding of all things. What is imagined one second becomes the inspiration towards what is possible the next. '42' deals with evidential truths, those ideas that can be tested or deduced by everyone. It is also about what is *not known,* but can be theorised and imagined as plausible. I have sometimes used passages of fiction to expand on conjecture and make imaginative models of our universe more vivid. You can wonder about these illustrations and bolt them on to your own views. So, if you would prefer to believe in 'ifs' and 'maybes', at least mine will offer something more reasonable and contemporary than the faded gospels, and blood-splattered science papers strewn across the pitted battleground of intellect at war with emotional need.

I think no one should accept the truth about anything until all truth has been wrestled from nature. My work is not asking anyone to close any doors on seeking alternative answers to their questions. We can all be wrong, or convey something inaccurate by mistake. I am not God, or his son, or the devout self-righteous sage who would say, "This is it... the truth... forever and ever, Amen!" *My truth* is extracted from what is understood about our universe today. Whatever may be constructed by discovering additional truths tomorrow, next year, or a million from now, I leave for others to consider and debate. I am hopeful this work may help to enlighten human thought, and bring about a greater degree of harmony between each struggling member of the human race. It would be a bonus if, after reading this, more people found a common direction through life instead of the diverse, and often conflicting, avenues we currently travel to make sense of it all.

What I have to offer here is a heuristic work! I am attempting to draw a series of pictures using fact and fiction in understanding a problem and exploring an answer. My methods are designed to expose the absurdities contained in resolutions offered by religious and scientific theories to the question of the meaning of life. My intention is to reveal my working out so that it stimulates further investigation by you, the reader. I believe no matter which course you choose other than mine to answer your own self-searching questions, my answers can be adapted to your own beliefs even if we differ slightly on the details and mechanism of how we derive a solution.

Some of this work examines the more local problems we experience in the days of our lives. How should we live a constructive and worthwhile period on this planet? How do we advance happiness, both our own and that of our kindred? If I have done my job correctly, we should be able to explore some interesting insights together, and find sustainable answers to these questions. I have deliberately avoided writing scientific explanations in jargon, and instead I have tried to make many pioneering concepts understandable to everyone. People already versed in the state of knowledge at the various frontiers of scientific investigation will recognise my statements are true (as far as we know right now). Anyone else less informed about science, and the theories underpinning current trends of investigation, will need to put a little faith in me. I have annotated some of the more critical statements in the major sections of this work, and I have provided further clarification at the end, together with suggested published works by other authors, for anyone curious enough to read up and check my facts.

I wish to make two clarifications now; they are both important in understanding my references to consciousness and to the non-aware state of the universe. When I talk about being 'aware', I have often used the words—conscious, consciousness, self-aware, aware, sentient, thinking, and possibly one or two other terms as well. Each of these expressions has slightly different meanings to a perfectionist, but I have used all of them synonymously as a state equal or higher to human consciousness: the feeling we have of knowing we are conscious and aware.

When talking about the universe, sometimes I am really referring to everything in it apart from us and, at other times, everything including us. One should regard the universe as the totality of everything in it. I also infer it to be a mathematical universe although it is more accurately defined as a program-universe. My concern here is to distinguish it apart from something having any will, or human-like purpose, and to avoid inferring any direct intelligence, intent, mind, or perception to it. As far as science and I are concerned, the universe is not alive and does not possess a mind. All its activity, including the existence of living things, is a consequence of properties, traits, and processes inherently woven into its constitution. The cosmos behaves more like a few lines of recursive software code, bound by mathematics and rules, which—along with trial and error—blindly gives a physical existence to the unknown abstracts driving it.

I would like to explain one final important point before you run out of patience with me. Most of us mere mortals find it difficult to come to terms with our individual ultimate deaths. Many readers will have strong belief systems. However erroneous and misinformed these may be from the truth of all truths, some kind of faith enables millions of people to come to terms with finality. A life without hope and magic is like a grey morning where the sun never breaks through for an instant. If you need to believe in magic to confront death, despair, and misery, then a beautiful message is already encapsulated within this work. What is important to me is that you find the message, think about it, and come to understand its astonishing implications. Afterwards, you may genuinely feel you are a worthy and important living entity because of it. The degree of belief required to do this is much smaller than any you currently apply to follow the particular faith you already have. This one is based on looking at what is here now in this life and this world, not on the promise of a non-evidential one in the hereafter. *Whether you know it or not, I want you to realise that you, and everyone around you, are critical to the survival of everything there is and will ever be.*

Chapter 1: Seeing clearly

Me

When you pick up a book, either it is because you wish to be entertained or you wish to learn something. The best book, be it fiction or fact, does both. I think any form of communication should use simple language to explain the most complex things. An author should be alive in the book—not for herself—but for the reader. If you are going to spend time listening silently to the thoughts of someone else, why be bored?

I think mostly about why I am here. I decided way back in the past, when I walked in my childhood places, that life was a strange and unknown thing. As I grew up and encountered this world, I decided I would continuously learn to understand what purpose, if any, there was in being alive. I have always felt a spiritual purpose, which is completely incongruous to all the established faiths. What I feel is an innate sense, something very different from the preaching and sermon of priest and Pope. Religions are human-made with no bearing or valid association with the real sense we all share of there being something more to life than is obvious to us. I have always wanted to write about reality, life and death, and to share my thoughts about these things with others. I thought I would write this book twenty years ago when I was much younger and had a more vibrant and imaginative mind. The problem with suspecting a facet of the truth, and being able to know it well enough to communicate it effectively to others, requires a great degree of experience. Now, at an autumnal age of fifty-six, I seem to have worked across so many different skill areas, including communication and technology, corporate-presentation, writing, video-film production, software programming; the visual arts, science writing, and self-publishing, that I believe I can accomplish this one task successfully. Many works similar to mine are aimed only at those people in society already knowledgeable of science, and able to understand quite difficult terms and theories easily. My work is intended to be accessible to all social groups, and be interesting to people who normally find debate about things like space, religion, philosophy, and atomic theory a bit tedious.

Scholars, people duly educated through university and achieving significant and worthy qualification to win public credibility when expressing their ideas, are normally the right people to produce books in the realm of science, theology, and philosophy. I am worthy to speak to you by no such avenue. In fact, I am completely and informally self-educated. Rather than see this as a disadvantage, I consider my more liberal route towards learning, locates me in a uniquely privileged position to communicate effectively with you, the reader. I have no institutionalised position of academia or professional reputation to defend. I am not conditioned by the narrow processes involved with presenting knowledge to examiners in order to prove I can grasp, and remember again, what was duly taught to me. Neither have I been 'channelled' by career aspirations into a single discipline of learning; instead of becoming an expert in one area of knowledge, after fifty-six years of constant probing and discovery, I have become more than proficient in many.

I am free to self-determine truth from fiction, bias from unprejudiced notion. I have checked my arguments and facts against those presented by formally educated authors and their works. I have seen errors of judgment in the diametrically opposed arguments between theologians and scientists about the nature of life, universe, and God. Each party, when presenting evidence to bring their arguments to the reader, has fallen foul of an unenviable trap: scholars and experts are also defending their entire area of study, their peers, the institutions they are members of, along with their professional reputations and protection of financial income. Their arguments then are nearly always unwittingly contaminated with a high degree of self-interest and protectionism.

My income does not depend on the success of this book, and my reputation is this field or that is irrelevant. The fact you are reading it is because my work was deemed well written, non-libellous, pertinent, and sufficiently worthy to add to the exploration of self-understanding. Publishers have risked large sums of money to produce and distribute this book. They are not likely to do that if they thought I had written a load of nonsense. My qualification is earned not by a phased sequence of tutorial course and examination over five or ten years, but by a lifetime of constant and deliberate learning in order to accomplish one thing: to deliver the culmination of this learning to others.

Who are my examiners?

You are!

It is important you recognise this book for what it offers—definitive answers for some of the most profound questions ever asked and left unsatisfactorily answered. Who I am is irrelevant, except to endorse here my qualification for producing it. The world is full of crackpots spouting ideas in the hope of personal fame and fortune, or who are simply self-deluded by their imagination and fantasy. Wishful thinking, speculative theories, romantic desire for magic to exist in a harsh and dangerous place— all have no justification from which to draw truth. These are human devices used to remove terror and foreboding from the realisation that our lives are limited, and the depressing thought that each of us, ultimately, will no longer exist. Religious belief systems, as they have evolved into what they represent today, look glaringly obvious to be elaborate daydreams: fragments of the same denial. What you and I are to discover together in these pages with equal degrees of cynicism, wonder, intelligence, observation, and conviction is a totally new non-belief system. It is not a figment of imagination aimed at removing the fear of dying.

I am betting all my time in writing this book that you and I will discover the only answer to all the things you have ever asked will end in a single answer, my answer, teasingly called—42: a concept which can be argued and challenged, but not disproved. The coming of increased intellectual capability through advancing technology should increase its value and prove its worth in time. The world does not need a new religion, nor does it find the dogma of science a complete solution to our human condition. Something new is required: truth, reason, and a unifying idea to unite everyone in experiencing life with global purpose and meaning as well as local ones. We need a novel and fresh approach to understanding our place in nature, one benefiting all conscious entities, be they made of flesh, machine, silicon chips, or hybrids of these, and one so convincingly obvious, it can be carried by our descendents into an ever-expanding and rewarding future.

Us

In the wonderful satirical story Hitchhikers Guide to the Galaxy by the now dead but not forgotten Douglas Adams, a planet's inhabitants spend thousands of years constructing a super-computer to ask it '*The answer to Life, the Universe, and Everything.*' The big day dawns when everyone waits to hear the answer after the computer has finally processed the data. The question is asked: "Oh... supreme computer... what is the answer to the meaning of life, the universe and everything?"

A great silence falls across the multitude waiting for an amazing revelation as the seconds tick by.

"42!" the computer announces proudly in a sombre voice.

Here lies the problem! If we don't understand the answer, maybe it is because we never asked the right question, or went through the working out ourselves. Remember doing mathematics at school? You laid out the sum, showed your working-out clearly, and even if you made a mistake in the final answer, you normally still got good marks. The reason for receiving them when you got the answer wrong is that you proved you understood the question, and gave evidence of how you worked through the problem to derive a solution. That alone is the real answer—not the digits on the bottom line!

Life is a bit like that: your life, my life, and all life. In my work here, I want to take you to the mind-boggling frontiers of science without 'boggling', and I want to take you down to the basics of everyday life without telling you, in advance, where we are going. When I talk with people about the stuff of life, I flit from one topic to another, because it seems to be the best way to make connections from diverse ideas, and bring them all together to create greater insight. I once made what would have been a very boring presentation into an interesting one by taking the risk of doing something very similar. I assembled a collection of ideas together, each from a completely different direction. When presented in isolation, each element had very little impact, but when all the bits finally came together— doubt and suspicion in my audience were replaced by surprise, delight, and realisation. On reflection, I think this two-hour presentation, given to senior managers of one of the largest corporation's in the UK, was the riskiest one I ever attempted. The methods I used had rarely been tried before, and were very unorthodox, but the result earned me a standing ovation and a hastily purchased bottle of champagne from an appreciative and delighted audience. Therefore, I would like to attempt a similar thing here, and ask your trust to go with me in this direction or that. At times, it might seem I am being inappropriate or informal, emotional instead of rational. Is this not the way of humans? I will often lead you into some fictional accounts, instead of factual ones, as a way of using example, metaphor, and bold imagination, to aid the truth when conjuring real vision and insight into view. If you find me criticising either religion or science in this section or that, or ridiculing something you passionately believe in, please stay with me and be patient; you may be pleasantly surprised by my shifting, and often, contradictory stances as I unravel deep-rooted ideas, and expose the truth buried beneath a heap of false beliefs.

Since this book is about the purpose of life, much of it is about discovering the nonsense already fed into our brains by people with intentions not based on making life an enjoyable and significant experience. The past centuries have been over-populated by people, in various positions of advantage and power, who were not intent on the idea of illuminating truth to others. Instead, they were focused upon maintaining their own advantage over the less informed. Many of the educated, and often self-important, scholars, priests, and cultural councillors of the past have enjoyed rewarding lives by sacrificing the well-being of the very men and women they were supposed to be guiding and protecting.

It is the 21st century and nothing has changed!

If you would like proof of this in recent times, consider the notion of slavery. Are we slaves today? I know we don't have whips cracking at our backs, nor are we in chains like our stereotype images of past slaves. But do you go to work each day to perform a task where you receive less than full payment in return? Does anyone *not* go to work with you in your company, but still receives reward from the work you do, instead of them? Big clue here: is a shareholder paid a financial dividend?

Do you live in a house purchased with borrowed money? If you live in the UK, or anywhere else in the western hemisphere, you probably do. When you go to work and earn money to pay a bank or mortgage lender for *their kindness* in providing you with finance for a home, you are doing nothing different from what your ancestors did in the early Middle Ages. Most common people went to work on the land of a local lord, growing food, and paying back the landowner some of (most of) the crop in return for the opportunity. Incidentally, this land was probably taken from these very people, initially, through an act of aggression and war.

Lords never went to farm the land. They never needed to: people like us were doing it for them! Is slavery abolished or is it still with us today, cleverly sanitised for acceptability in the 21st century? Where the sword once kept us in our place, today the new lords do it through the power of the media first, and a dogmatic, clinical, and very expensive legal system second. If you fail to get the media message, and drop out of the system or rebel against unjust laws instead, there is nowhere to go but down and out!

The Power of Deception and False Beliefs

I believe we all suffer from some degree of false belief because of misinformation, ignorance, non-acceptance of truth, bias, arrogance, or any number of other root causes. We are human. We are made this way. Religious belief is clearly acknowledged to be just that—belief, but few people realise *scientific beliefs* exist too. The frontiers of scientific exploration are bordered with unseen fences dividing the known from the seemingly impenetrable no-man's land of the unknown; pioneering minds often attempt crossing new ground, risking reputation-mine-fields, armed with no more than sketchy maps, and their individual beliefs that they are on the right track. I think it is a good idea to expose just how vulnerable we all are to believing in things in our local environment, which, for the most part, are not proved true, before approaching the gigantic task of probing, and removing, many of our existing beliefs regarding the meaning of life. If we are to peer further into the depths of the greatest mystery of all, we need to do so with clear vision, and without blinkers glued to our eyes to make us feel comfortable with a perceived reality instead of the true one. What better way of doing this than looking at common ideas we believe in today. I wish to make clear I am not doing this with any political purpose. I am attempting to demonstrate how readily we are all fooled by propaganda, and the reiteration of bits of information about issues, which we have little opportunity to examine in fine detail. I hope by doing this, it will help reveal the deception of our minds by the more impassioned, but equally misguided, ideas governing our view of reality, and our unique role within it.

Much of what we do as a populace is cleverly controlled by other people who realise they are doing exactly that—controlling and influencing us. We are coerced into believing happiness is dependent on obtaining material wealth, and avoiding threat. We are conditioned into being forever fearful of the next bad news item: train crash kills twelve, suicide bomber kills twenty, bird-flu to wipe out 20%, killer bees spread to USA; serial rapist strikes again, national drought forecast for summer… Most of the news items pouring from our televisions, radios, and newspapers seem to centre on five major topics a day. These are the ones we are obliged to carry around for the next twenty-four hours, sometimes longer, until the next five are announced. Maybe you haven't noticed this five-a-day newsy thing. Test it out for yourself. Meanwhile, across the globe and in our own country, good events are happening too at a rate equal to, or greater than, the bad ones. There are always other things going on, which often have more serious consequences on our lives than those reported to us. I think all the news channels must be buying the same news from the same source (Reuters?) or maybe news-reportage is being invisibly managed. What we hear, see, and are *allowed* to witness, has a major influence on our beliefs, and the way we focus on the world we live in.

Many false beliefs control or influence our thinking right now. Let me demonstrate how misinformation influences our notion of truth. How many of the statements below do you think are true?

- *If you save money all your life, when you get old and cannot work, your savings will protect you.*
- *If you go to church, or behave as a good person, God will love you and give you a place in heaven.*
- *A global plague will evolve anytime now and bring worldwide death: bird-flu, AIDS, Malaria, and TB.*
- *Western governmental systems are democratic.*
- *Corporate monopolies do not exist in the United Kingdom.*
- *Global warming is man-made.*
- *The USA is a Christian country.*
- *The Holy Bible was written to be the foundation of Christianity.*
- *Young people create the most graffiti in our public places.*
- *Anarchy means revolution and chaos.*
- *Income tax in England was brought about to fund the needs of society as a whole.*
- *The greatest threat to you right now is terrorism.*
- *A police force was created to protect the population of a city or country.*
- *CCTV cameras protect you.*

I could suggest more but I won't. What I will say is—all of the statements above may appear true or false based on your belief in them or lack of it. There is no conclusive proof or irrefutable evidence that any of them are true. In fact, most of them, if not all, can be said to be predominately false! Let me show you some examples:

A police force was created to protect the population of a city or country.

It was not! The police force in England was created as a solution to protecting wealthy people when entering the East End of London, normally 'gentlemen', looking for women who were prostituting themselves because of impoverishment and lack of education. These wealthy people paid the Bow Street Runners for their protection services.

CCTV cameras protect you.

Oh no, they don't! CCTV cameras detect and report on a crime after it has taken place. People who wish to mug or murder you can avoid detection by wearing hoods, masks, and the like. The cameras may well help the authorities determine who murdered you on the train platform, but they are not going to prevent it from happening. The cameras will only help 'crime-solving' statistics, and assist in monitoring citizens who the authorities (both malevolent regimes as well as benevolent ones) deem to be potential threats to *whatever they also deem* is important to protect. Most of the time, cameras are not there to protect you, but to catch who harms you, or to detect enemies of the establishment. George Orwell must be screaming, "Watch out! I told you so!" from his grave.

Anarchy means revolution and chaos

It doesn't. The word 'anarchy' has been bent to represent a negative state by those people who consider a population is best managed by control and regulation, often members of a society who *know* what is best for us. Anarchy does not mean the total absence of rules, but more the ideal state of an anti-authoritarian society based on spontaneous order of free individuals in non-led communities—people operating on principles of mutual aid, voluntary association, and direct action. Anarchists believe all people are imbued with a sort of common sense, which allows people to come together in the absence of any government, and via mutual agreement—form a functional existence. Anarchy does not reject ethics or principles but rather 'imposed morality' (visit Mykanos, the Greek island to see a good modern example).

Global warming is man-made.

Is it? Global Warming and Global Cooling have happened many times before today in constant cycles since the formation of the planet. The Romans grew grapes and made wine in England. Greenland was so named because of the vast expanses of lush grass observed across its terrain. Humankind may well be speeding up a mechanism that already exists, but there is no conclusive proof of our presence, or our activities, triggering any event, which would not be occurring already if we did not exist. Scientists obtained the first unequivocal evidence of a continuing moderate natural climate cycle in the 1980s, when Willi Dansgaard of Denmark and Hans Oeschger of Switzerland first saw two mile-long ice cores from Greenland representing 250,000 years of Earth's frozen, layered climate history. From their initial examination, Dansgaard and Oeschger estimated the smaller temperature cycles at 2,550 years.

As a footnote: our most sophisticated computers are used to model future weather and climate possibilities, based on data recorded now. But even our best climate modelling tools are unable to predict weather conditions for more than a few days ahead; the number of variables involved, and the complexity produced by even a small change in one them, make it impossible to calculate accurately anything useful beyond a small time frame. We don't even know if globing warming is likely to produce more rain, or less rain!

You can see what I am saying, I hope, without my needing to overstate the point. If you are interested in the deceptions and falsehoods contained in the other statements, I have provided examples of how they can all be challenged in an appendix at the end of this book.

We all have beliefs. Most of them are not actually founded on any truth whatsoever, just snippets of information, half-truths, incomplete facts, and the management of information by people who benefit the most by duping us into supporting false belief. The beauty of 42 is that *it is not* a belief system; everything is deduced from what science currently understands about reality. Truth will out itself, and stand boldly in the face of lies and falsehood. It can be wedded to, or replace, any belief system in the world, because it will challenge whatever people believe in and offer them choices. It is up to you. Carry on believing in whatever you were told before, or come to understand the reason you are here, your one true purpose—and what you can do to help that. What's more, you can enjoy your life doing it, experiencing the pleasure of knowing what you are doing is good, and that it doesn't involve regular jaunts to expensive church buildings, sitting in confessional boxes, or dragging a prayer mat around with you everywhere.

Let's begin to see the truth by starting here at the point where life and the universe really began…

42—The Answer to Life, the Universe and Everything

Chapter 2: Ant Minds

Just how did the universe begin? This is a profound question. Most people have heard of the 'Big Bang' theory.[1] The concept is once there was just nothing; no universe, no air, no space, no vacuum. No existence! Suddenly, a microscopic amount of substance, call it energy or matter—'something' *came into existence* and then exploded at an unimaginable rate, speed, and vigour: the *Big Bang!* Since all the stars, planets, and rocks in the universe stemmed from this one moment, the density of the original microscopic mass would have been so extremely high, it's a wonder it didn't as quickly turn into a super black hole and disappear again. But it didn't, and consequently our universe came into existence— followed some 13.7 billion years later by us!

There is a lot of evidence to support many strands of this theory, although perceptive minds, like that of Stephen Hawking, have recently brought into focus more advanced models for what may have happened involving a term called Branes[2]. Much of the new theory involves our time and space emerging from the interaction of elements existing outside our perceived reality and universe, in higher dimensions, and really describes causation rather than effect. In essence, scientists currently determine that our universe did begin from a single point, arriving from a 'nowhere' into an existence *it*—this something—constructed for itself. The microscopic mass then expanded/inflated in just a few microseconds, growing to a phenomenal size (it still continues to inflate today) and establishing a 'somewhere', which we describe as the Universe (or reality) that we exist in today. Interestingly, at least one holy book, the bible, shares a common notion about creation, but expresses it subjectively as an act of God. If one removes the subjective and literal elements from the story of Genesis, the text, which describes the sequence of events about the emergence of the Cosmos, is uncannily accurate!

Genesis*: And God said, "Let there be light: and there was light."*

If intelligence continues to endure, be it human or otherwise, then a more refined understanding of the universe, and everything in it, will ultimately be teased out of ignorance. Whatever the ultimate truth is, I bet it makes little difference to the daily life of any humans living later. For all we know, what we experience as life may just be an illusion—a fantasy game of our real selves in another place and time. The problem is we still have to live it—this life—irrespective of its true nature! Knowing the answers to ultimate questions will have an impact on our overall outlook on life. We are designed to anticipate and survive whatever is coming next in our practical lives, and this occupies our minds more than thoughts about reality. Therefore, unless the method of creation has a sinister by-product relating to our immediate or ultimate demise, knowing how it all began, is not going to make us any happier or more successful living through the details of daily routine.

Whatever the mystery of creation turns out to be, only a super-aware entity, thousands of times more advanced than us, will ever really comprehend it. Each time I discover an ant in my kitchen and gently pick it up to explain to it where it is, the silly thing doesn't seem to understand. Yes, yes... I know ants can't speak English, but even if I were able to communicate with them using antenna vibrations and chemical messengers, there is little likelihood I will be able to map my internalised view of the kitchen, along with its position in the grander scale of things, into the tiny neural network of an ant: a quart will not fit into a pint glass! Only our technology will ever have a realistic opportunity to comprehend the universe's creation fully. We may be doomed forever to be no more than ants—tiny minds, grasping only simplistic representations of the advanced understanding communicated to us from our future super computers. I believe the only way we will ever understand the answer to everything is through expanding our personal and collective awareness.

Many people believe God made the universe. Not just any god mind you, but the supernatural one—you know, the old white guy with the long beard, sitting in a place called heaven: *a character with all the attributes we possess.* This god, the one God, is so perfect, He—according to tiny ant-priests, who for nearly two thousand years, have vigorously supported the written details of several monumental fables—is the one who judges *us*. According to the Christian viewpoint, our one god (just like humans) gets angry. In fact, he can be so vindictive that he is quite willing to let the Devil have us at the slightest sign of disbelief in Him. I am amazed at how He could knowingly abandon each one of us to an eternity of misery, torment, and pain in the inferno of hell just because we are human and misunderstand things. Surely, if God is so emotional and prepared to punish his children so absolutely for one error, which I see as *his* fault for not making clear to us, why leave the message in the hands of ambassadors just like us humans? Why not instead pop in occasionally on a grand scale to deliver the message in person. Even the President of the USA, and the Prime Minister of England, take time for a walk among the voters to gain their confidence. Nice one God!

I may appear flippant about this picture of God. Worse still—irreverent! In light of 21st century thinking, perhaps we need to separate the traditional religious perceptions of God from the religions themselves, to understand how outdated the models of faith-gods are. We have come a long way in our comprehension of reality through scientific thought, and our fresh insights have inspired many people to abandon this simplistic picture of the supreme deity. Unfortunately, even our updated concept of a supreme being is still contaminated by thousands of years of conditioning. We like to believe He exists, but many of us understand this cannot be in the literal way described in the bible and Holy Scriptures.

I believe it's important to deconstruct this model of a religion-based god, and to challenge how it can still be considered a viable idea. Please tolerate my flippancy a while longer, as I examine what the Christian church has historically tried to define as the creator of the universe. We are informed that God does not really wish to lose us to the Devil, but hey… he gave us free will, so if we don't behave, it is entirely our fault. It appears to me that an egotistical god, demanding our worship and our free will to side with him, is more likely the construction of men on earth desiring our loyalties to *their* earthly ambitions instead—or else suffer the consequences. A quick visit to the land registry office to see what is owned by religious organisations, shows how effective this fable has been in establishing wealth, position, and power for the benefit of those still telling us stories. It has been said that the Vatican has huge investments with the Hambros Bank, Rothschild, the Credit Suisse in London and Zurich; has substantial investments with the Morgan Bank, the Chase-Manhattan Bank, the First National Bank of New York, and others. It is also said the Vatican has billions of shares in the most powerful corporations in the world: Gulf Oil Shell, General Motors, Bethlehem Steel, General Electric, etc. Religion is the best business in the world!

Just to ensure I am not guilty here of tarring all priests and holy workers with the same brush, I understand quite well how many people, wishing to do good, join religious institutions as a means of having opportunities to do exactly that. I am sure many of these folk do not fully support all aspects of their chosen institution, or its interpretation of how and why good things should be done. For them, a church-based regime offers financial support, and a way of life to enable them to deliver their personal good; they suffer the rest of religious ritual and dogma in silence!

The second largest belief system in the western world, Islam, has a god who seemingly does not consider women equal to men. In fact, within deeply devout Islamic states, religion and politics are one! These are the countries where if an Islamic woman makes a small human error of judgment, and copulates with a male outside marriage, Islam's god believes a fitting reaction is to abandon his creation, and leave her at the mercy of rock-throwing crowds until she is bludgeoned to death. It sounds to me like the same vindictive god of Christianity, or more aptly put, the same vindictive and insecure men demanding the abeyance of *inferior* women to *their* will: not God! I will go a step further, and suggest this god, the one true God who made the universe with his mighty mind, is more the construction of self-made *men* wanting nothing more than complete servitude and loyalty from *their* lowly subjects. I suggest each of our gods is nothing more than a bogey-man, fabricated, and kept alive by those people on earth (men?) who have the most to gain from oppressing the rest of us mortals with such formidable tales; this supernatural god is no god at all, but the clever tool by which priests dominate societies with a confused mix of religious and cultural ideology to keep themselves dominant and powerful!

There is also another god. Not a supernatural one. It is the god often referred to by people of logic and reason—scientists. Their god is, in fact, not a god either, but a label representing the natural universe as a collective set of rules and laws. Their god is not sentient. It is a mathematical set of ordered, natural interactions—unaware of any 'self'. Without a degree of reason, we humans have no capability other than the same as worms, ants, and slugs. Reason suggests that evidence of a phenomenon, which can be examined by others, provides truth. Knowing the truth about anything provides an advantage to any creature having sufficient awareness to exploit it. So, if you and I are driving along a similar route from London to Oxford for example, which one of us is most likely to make the end of our journey safely: the person who believes he will reach his destination if he keeps driving along any road in the rough direction of Oxford, or the person carrying a map, because someone else undertook the journey first, and provided the evidence for another to follow? It appears to me that faith, without any supporting hard proof, is a recipe for ruin and disaster!

Evidence of a god, creating everything, does not exist anywhere. It does not mean there is no God; it may take a long time to discover such evidence if a one does exist. The problem with all religions is we are asked to *believe* in a god just through belief itself. Why? Where is the evidence to convince us? One common answer in a modern world is now muted to be 'Intelligent Design'. The idea behind this sinister term is the universe is so complex, as are living things, that an intelligence must have designed it—God!

Really? Actually, beautiful and profound complexity is generated by the smallest set of simple rules. If you really wish to understand this, look at John Conway's 'Game of Life' at *http://www.bitstorm.org/gameoflife/*

This is an example of 'cellular automaton' invented by Cambridge mathematician John Conway. It became widely known when it was mentioned in an article published by Scientific American in 1970. The game consists of a computer program where a collection of cells is based on a few mathematical rules. They live, die, or multiply depending on the initial conditions, and form various patterns throughout the course of the game. The rules are as basic as any can be, but the emerging patterns, and activity, are fascinating to see.

There were many gods before the Christian faith was adopted as the one true religion for the entire Roman Empire by the Roman Emperor Constantine, about 400 AD. In fact, humankind believed in a multi-god faith for thousands of years, far longer than a single-god one. Maybe the idea, of belief in a single god, provided a brilliant opportunity for Constantine to strengthen his position. Subjects worshipping a variety of different deities are less likely to associate an emperor with a single ruler of the known world. People within an empire, looking at one god for their salvation, are far more likely to accommodate the idea of a single earthly ruler—Constantine!

More pertinent today, why should *one* true god be the creator of something as awesome as the universe? There is a monstrous absurdity with this argument. Computer chips are extremely complex in their wonderful designs. They are so intricate today, they can no longer be created without a team of people and computers working together to determine their layout. If Intelligent Design of the universe is indicative of it being made by an 'awareness' superior to ours, then existing examples of sophisticated design on Earth, suggests a team did it, and not one all-encompassing intellect. Wherever there is an example of us making complex objects, we discover an increasing dependence on teams of minds, and technological intelligence to accomplish the task. A far more likely god would be a computer. But this idea would certainly not fit in with Christian theology or the Intelligent Design argument. As with all antiquated belief systems, designed to resolve the ultimate question of why we are here, the struggle today for their champions is to find some way of making their absurd ideas fit alongside the often-ponderous, but testable, reasoning of science. And the myth goes on forever! Only when we remove religious instruction from all our schools, will our descendents have a clear choice at examining what is true and what is false. If a belief in why we are here is so important, it can surely be taught on any given Sunday, full time, and in the priests and parents time using the resources of those who wish to continue filling young and impressionable minds with ridiculous fiction.

The world is now made up of many countries inhabited by multicultural societies, where each sub-culture has a different belief system. If people are to proceed in peace, and tolerate other people's beliefs, then it is better to divide culture, and way of life, from religious ideas. We need to share a common set of rules based on social order, and a common education based on fact and truth—not myth and fable. What a progressive society needs to do, as soon as possible, is to accept the fact that many of its members are unwitting fanatics, and then shift all activities encouraging the spread of their madness to other people, away from mainstream education. All schools, in all countries wishing to rediscover social adhesion, should be made non-faith institutions; this way, at least one major division in society—caused by the legacy of inflexible closed-systems, irrevocably binding culture, belief, and politic—would disappear in a single generation.

Observation from the philosophy of 42: There is no hard testable evidence of a God, so we are asked to act without reason and just believe there is one anyway.

Chapter 3: Super Minds

A scientific view of how the universe was created is in a constant state of review. Leaving aside the creation of the universe for one moment—how did *we* come about. Two of the many natural elements in the universe are carbon and silicon. All life on this planet is founded on the material we have come to name as carbon, which, at an atomic and molecular level, has a wide degree of flexibility in how it can combine with other natural elements. Good evidence shows one testable reason we are here is that these two elements combine readily with other materials to make sophisticated and complex compounds. All life on Earth is here due to these properties in carbon.

Science, in stark contrast to other *non-scientific* doctrines speculating about a mechanism for the emergence of life, advocates the reason we are here is that we are the product of natural laws governing the way a universe evolves. One inherent property of the universe is very interesting, and I believe it can be considered one the main reasons why life emerges in our universe; it is the internal mechanism of energy-conservation and its by-product: *entropy!* When energy is being transferred—for example, by rubbing two sticks together to produce heat—some of the energy is used effectively, but some of it will also dissipate and be lost! Energy can only be transferred in one direction: from potential energy in a low-entropy and ordered equilibrium state, to energy-release, and the realisation of its potential to make something happen—activity! Whenever energy is coaxed out from its bounded and natural resting place in matter, to initiate and drive physical activity, some of the energy released will be lost forever, and cannot be restored. All activity in the universe: suns burning, marbles rolling, kettles boiling, hands rubbing... is ultimately fuelled by energy-exchange accompanied by a loss of energy. It will take a very long time but all those packets of energy being lost will finally add up to one thing—the total loss of all useful energy to make anything happen ever again. No energy, no more activity, means no active universe!

Living organisms are the only structures ever discovered with a unique property: *they defeat death!* Suns burn out and explode. Rocks and planets grow colder, and become inert, as they lose the original heat that spawned them. The entire universe will continue expanding and cooling until it is dead—cold and fragmented across a space and time so vast, a single atom has no chance in all eternity of ever combining again with another.

Living things also experience entropy: we grow old and die! But remarkably, life contains a special trait not observed in non-living things. Organic forms are the only known structures able to defeat death caused by entropy. How? Because they can replicate themselves before they die! This unique quality provides evidence of a mechanism, ordered by rules, and existing as an inherent characteristic within the universe's own structure, which can counteract the negative effects of entropy. Surely, a process so effective at combating the very assassin—entropy—destined to destroy an entire universe, is a remarkable and significant creation. The universe would be in absolute joy if it had a mind to know this. Of course, there is a catch: although living forms can fight back against the growing disorder in their internal systems (ageing) through the process of rebirth, they are dependent on the external world to refuel their own energy stores continuously. Life needs the resources provided by components of the universe to survive. No universe to draw energy from means no us!

I will go further and suggest only life offers any potential for a universe to defeat its own fated death. And boy, the universe is playing a long shot with that one, but *we* may be the only thing *it* (the universe) possesses to escape the inevitable!

Admittedly, life per se on this planet cannot even begin to save a universe right now. We are hardly in a position to save a planet at this point in our evolution. However, life is the only structure in the universe yet discovered with the potential for rapidly evolving a much higher degree of awareness. Rocks or plants do not seem to think. Thought seems to rely to all extent on the presence of a cell network designed specifically for the thinking process: a brain! Humans have brains. The enigmatic trait of 'self-awareness' is only apparent where brains have evolved sufficient complexity, and quantity, of inter-brain-cell connections (dendrites) to raise the question of self-awareness and abstract thought. When it comes down to consciousness in living structures, the greater the number of connections in the smallest space, the greater the opportunity seems to exist for quicker thought and improved awareness.

At first sight, our human brains seem to be capable already of a high degree of awareness. It is a shame most of us fail to employ this potential fully. Yet we have no yardstick to measure our degree of awareness against superior ones, and there is no evidence to suggest we are increasing our biological awareness, despite the recent development of ever more sophisticated technologies. I suspect human consciousness is only a tiny intellect. We can only perceive everything in three dimensions. It is easy to conjure up an imagined image of a 3D cube. Now try a 4D one! Zonk! Blanko!

Our computers, working with applied and specialised branches of mathematics, have no trouble whatsoever dealing with four dimensions, or five or six, but no matter how hard *we* try, it is impossible for *us* to model such abstracts onto our three-dimensional neural networks (brains). The mechanism of the universe, which teased us out of gas, energy, and matter, could only go so far in bringing about any kind of perceptive entity as a starting point. We appear to be it! However, our minds are severely limited by the inability to perceive and understand a universe existing, probably not in three dimensions, but more—ten!

Although human intelligence seems to possess only 3/10 (a sloppy guesstimate) of the required awareness to understand fully what everything is all about, it has sufficient potential to develop and organise other bits of matter into an increasingly '*more aware*' arrangement. The universe, through its own natural mechanisms, has developed living entities. When those entities discover information about the nature of the universe—initially locally (this field, that tree...) and then later globally (electricity, magnetism, atomic energy...)—they are able to use their awareness dynamically to influence the organisation of the universe. At first, this would be at a much-localised level. When we build a car, it is difficult to imagine every single piece of material in it has been dug from the ground, extracted from the sea or atmosphere, and then re-organised through our ingenuity into a purposeful object. If the awareness of living things is sufficiently capable of reorganising matter into a predetermined arrangement, however crudely to start with, it is likely to stumble on the capability to recombine materials and energies to create *improved forms of awareness*, and not just objects, thus increasing the degree by which living things are able to influence their local environment. Ultimately, 'awareness' seems to be the only tool in the universe's tool kit by which it can possibly one day have any possibility of remaining in existence. Reality, the universe, could be either replicated, or manipulated, on a massive scale through intervention by living, or non-living, intelligent beings.

The only sign of a god being present in this situation is one indicating a god may be in the slow process of being created, but there is nothing in evidence of a god actually existing right now. I believe it is reasonable to deduce there is no god presiding over us. The only potential for a super-intelligent entity to evolve in our reality lies within us humans. We are the ones who are potentially able to *create* God—not the biblical supernatural deity, but something very akin to having god capability. *We* have the aptitude to *evolve* an all-knowing intellect! This is not something achievable by blind faith; only through intelligent exploration of the time and space we inherit, do we finally come to glimpse any real purpose in being here. It has only taken a mere 13.7 billion years for the universe's long-shot gamble to be suspected. I propose this: life, us, and any other intelligent living forms residing elsewhere in the cosmos, is the crude mirror by which the universe begins to perceive itself!

Scientists are convinced the universe will continue to inflate, or deflate, ultimately having equally serious consequences. We have insufficient knowledge to know if other natural mechanisms exist to influence either of these outcomes. For example, Black Holes exist at the centre of all galactic star systems and suck away matter (stars and galaxies) from this dimensional plane at alarming rates. The degree of energy involved, together with the many diverse locations where this is happening, makes it improbable that matter removed this way can ever be reorganised again to evolve (or replicate) another universe like ours. Black holes may prove to be additional dangerous mechanisms, leaking away all useful heat from our system.

What we do know is that human beings have enough mental capability to tease powerful truths out of their environment, and use their knowledge to re-combine the stuff of stars into sophisticated, purposeful, and powerful creations. We began to extend our 'awareness' potential the moment we first used crude tools to record data externally to our minds. A mark on a wall is a memory cell. A page in a book saves us the work of storing a lot of data in our brains. We have extended our memory storage capabilities a billion fold in the last 30 years! With the creation of calculating machines and the continuous evolution of processors, along with the computing technologies they lay at the heart of, we are now exponentially extending not only our memory capability, but our reasoning one as well.

The crude technological devices we employ now to combine our extended minds with our innate biological ones, will have evolved to provide seamless interfaces in just a few years. Technological implants are already well established in several disciplines of medicine, and we have begun the priceless journey of hooking up our nervous systems with our external hardware in pioneering ways.

Unless this enhanced evolution is abruptly stopped by war, disease, meteor-hit, volcano, or something else unforeseen, human awareness will inevitably evolve (or better said: expand!) through the integration of mind, brain, silicon, processor, and genetic engineering. We may have ant-brains right now but in a thousand years, it will be impossible to distinguish brain cell and dendrite from silicon etched cell and nano-switch. Humans are highly unlikely to remain entirely human for very much longer.

Natural evolution of a minute awareness into a major awareness is a time-expensive process for our mathematical universe to accomplish with its inherent mechanism of evolution, but the evidence in front of us today, scattered throughout every office and home, proves something extraordinary is happening. Once 'awareness' gathers knowledge and comprehension at a certain level, it becomes an extended tool of the universe's evolution: universe makes life, life evolves complex nervous systems to aid its survival, complex nervous systems evolve brains to centralise and organise nerve function, brain evolves new biological areas capable of processing information not directly required for nerve function alone (spare capacity); new brain areas develop and evolve insight and self-awareness, aware life forms extend their awareness capability through external devices, aware life forms combine seamlessly with their previously-external, extended, devices to expand awareness rapidly and logarithmically.

Final result: *ant-minds evolve into super-minds!*

Observation from the philosophy of 42: 'Awareness' is the only known tool by which a mathematical, unconscious universe has the potential to avoid ceasing to be.

Chapter 4: Birth

I guess one of your main reasons for purchasing this book is to find out the answer to the following question: *why am I here?* As consciousness develops, every growing child reaches a stage where it ponders this frightening question. There are more queries connected to it. Do I have a purpose in being alive? Why must I die? What happens when I die?

I thought of one question as a child that many other people never ask. This single question provides help in understanding more about the other ones. What did I ask? It was, "Where was I before I was born?"

This is the way I reasoned an answer, of sorts. Just before I was born, I was a lesser me still growing in my mother's womb. Before that, I was not yet combined: I was sperm, egg, gas in the air, mineral in the ground, water not yet fallen as rain. Somewhat poetic, I agree, but absolutely 100% true. All of these fragments of the universe apart—not one of them is aware! It might be argued that the only important ingredient waiting to become me is genetic code—sections in my father's sperm and others in my mother's egg. However, without all the other ingredients being available too, the only thing likely to happen is the creation of one dead me!

What I am, therefore, has always been. I am made from the dust of long extinct dinosaurs, the gases of dying suns, and the energy between atoms. Far more difficult to perceive: my atomic self is directly in contact with every other atom in the universe, no-matter how far away and no matter how tentative or weak that bond is. My atoms are glued to the very fabric we label as space-time, the same space-time extending out and away from the local section on which I am stuck on. It stretches across hundreds of thousands of light years. Out there at the most unimaginable distances, entire galaxies are stuck on my sheet, your sheet, and everyone's sheet. We are all stuck together! The difference between the *me now* and the *me before now* comes down to a single deduced truth: here at this precise moment of time and at this exact location, I am *organised* matter stuck on that space-time sheet. Before now, before life for me, I was *disorganised* matter! Perhaps more accurately put—I was less *well-organised* matter, since even stray atoms, dust, stars and gas are all being continuously rearranged or managed in one way or another by the rules imposed on it by the mathematical universe.

Up to this point, I think these things are quite easy to see and understand. What follows is a bit more difficult, so stay with me. A highly specialised and extremely vigorous piece of organised matter is deliberately kept separated into two parts. If the two halves are brought together, something almost impossible happens: all those bits of other stuff glued aimlessly and locally on the space-time sheet begin to be influenced by the merging of these two specialised lumps. What is more amazing is they are pulled across the sheet, or are attracted down or managed sideways towards the merged lump. No magic is going on here. My mother is feeding herself, picking up a lot of that stuff and putting it into her mouth or sucking it into her lungs. An awful lot of it is being delivered by other highly organised structures (parts of her body, veins, carbon-based flesh, tubes) towards... you guessed it—little old two-halves-together me!

Let us look together at an important event in each of our lives but do it in an imaginative way. Close your eyes for a moment and perceive blackness. The darkness of closed eyelids makes an excellent screen on which to project what you and I are going to put there. Now imagine millions of tiny round balls, all shapes, all sizes, populating that blackness. Give them a bit of random movement and make some of them coloured: white, red, orange, and yellow... the brighter the colour, the more active the ball! You can have some of them closer together in little groups, maybe moving around one other in a constant pattern. Some of the balls bounce off one another, some bump against others and merge into a single ball, duller in colour; other balls shatter on collision into ten or twenty small balls, and glow so brightly, that they only exist momentarily as highly excited sparks, before fading into darkness.

This picture represents an extremely over-simplified model of the local area of our space-time sheet: the fabric of space on which all material adheres to. Those little balls represent atoms, dust, molecules, and stuff at a tiny level, which we rarely see in such a novel way. If you watched long enough, you would start to see those little balls slowly moving further and further away from one another, and growing ever duller in colour. If we were able to watch for a very long time, most of the balls will fade and become black, because they will have moved so far apart that the only contact they will ever have with one another will be through the 'third-party' contact provided by the space-time sheet they are stuck to. The balls have not actually moved anywhere; they are still glued to exactly the same spot they were in when you started watching them. What has happened is through the process of time; inflation has caused the sheet itself to become super-stretched as a consequence to an ever-expanding universe. Entropy is happening!

Okay. Now bring everything back again to our starting point. All balls are happily dancing around each other; right there, millions of them. Now imagine an entirely new object moving in from the right. It looks something like a ball but, when you look closely, you see it is an object made up of many smaller balls whizzing rapidly in complex patterns around one another. As it moves across your vision, from right to left, other balls are pulled towards it and then pushed away again. The object is fervent, very active, and its internal movements are more intricate when compared to the systematic dancing of the normal balls. Enter stage left, another object like the first. It's behaving in a very similar way. They appear to be on a collision course. We are witnessing the two halves of my genetic code coming together. I have omitted the representative forms for both my parents here because their large, and highly organised, collection of fervent coloured balls would make the picture too complicated to see what is going to happen next.

Suddenly, the two exotic objects—my father's sperm, and my mother's egg—merge; flash, light, confusion, a ripple across space-time; all the other balls shudder! When the smoke clears (pun), there in the centre of all the coloured balls is a very complex, extremely active structure. It is pulling all the other balls towards it, and each one that touches it becomes assimilated into the central mass, increasing its size and evolving ever-greater complexity within its constantly shifting form.

I am born!

No sign of high-level entropy here yet, but it will come in time of course. What you have just imagined, but never seen before, is a more accurate image of me being created than if you had a web-camera inside my mother's womb at the time, which, incidentally, would have caused a bit of trouble between my father and you. If I am successful enough to reach around 13 years on from the moment you just witnessed, you would observe a much larger, and more sophisticated—but still extremely vigorous—me as a mass of whirling balls pulling in most of the other balls around me, and assimilating them. Where was I before I was born? I have always been here but, for most of the time, I was not organised in this shape. Instead, I was an earlier exotic object assimilating other balls (taking pieces of our universe, and adding them to my shape and form). Before the very first exotic shape, I was less exotic, and less organised; before that, I was made of scattered balls, bouncing against others, looking for appropriate carbon balls to merge with. At all these stages before now, I had no awareness; I was simply being pushed around by the innate mathematics of the universe. *It*, the universe, already contained a set of functions, like program code, to influence carbon balls, and combine them with other balls to make countless variations of semi-organised structures. One might ask why the universe has this encoded property, and the answer can only come down to one of three choices: chance, design, or recursion—the third option suggesting our universe is the result of a legacy from a previous one; or that it evolved from a reality already containing the seeds of the properties it has now, and they simply got passed on. Whatever the answer, ultimately in our slice of reality, a group of balls became combined through random trial and error in such a way they were unable to stop reorganising and assimilating other balls. Life emerges from the stuff of stars. What is important in all of this is *energy*. Other properties are required to keep the show on the road: gravity, time, space fabric, but a seemingly simple yet deceptively complex thing—heat, powers all activity throughout the entire universe. No heat: no activity! Most of the heat in the universe is still contained in the stars. Most of the observable matter in the universe is also contained in the stars. Once there were only stars. Where was I before I was born? I was in a star.

Astute people might try to say that what I have just described is not the sentient me being created at all. If I am only made of atoms, how would I be aware of me? The same people would say I have just described only the early forming of my body, including my brain, but not I, Mol, the *awareness* that evolved within that structure. They would argue I am unique and different from all other minds in all those other structures like me—other human beings! In only a very small way, I will agree with them slightly. However, if I am really *my* awareness, what am I: the awareness I was at two years old, twenty years old or seventy years old? Alternatively, am I just the awareness I am now: the product of all the data and experiences in my tortuous journey through life up to this point in time?

The answer is stunningly simple. None of these! Although I am self-aware, there is no 'I'. I think this is an elaborate illusion! It is important to learn how perception of individual self is actually an astonishing deception. You, me, Napoleon Bonaparte, Adolph Hitler, Jesus Christ, your mother, sister, brother, friend, and enemy are one single developing awareness strung across many physical containers. The reason we are fragmented and not together in one total consciousness provides advantages: (1) it is a good mechanism for avoiding total awareness-extinction; (2) one awareness, divided across different minds, is a more secure way of ensuring a variety of perceptions are able to consider reality, before a universe constructs an accurate self-portrait. The universe does not put all its proverbial eggs into one basket! This will become apparent shortly as we unravel real self from unreal self in the next chapter.

Chapter 5: The illusion of self

"I think, therefore I am," said philosopher René Descartes in the 17th century. If there are thoughts, there must be someone who thinks them. Alas, René had to ponder the question of awareness and human existence as a complete and inseparable issue. Seventeenth century science had not yet explored the mysteries of the brain, or discovered how electronic systems can simulate awareness artificially—albeit, the latter is still a young developing branch of science. Knowledge of sub-atomic particles, electricity, string theory, and quantum mechanics were all waiting to be found. The more we uncover about the nature of our universe, the more difficult it gets to pin down what consciousness is precisely, let alone 'self-awareness'. At sub-microscopic scales, the cellular make-up of living organs is constantly being replicated as material at an atomic level is replaced. The smallest discernible building blocks of living tissue—cells, are in a constant state of renewal, reactionary change, and repair. Is each of us the 'I am' in any arbitrarily chosen moment—the exact pattern of dendrite matrix (the tiny branches extending from and connecting with each brain cell), or are we each the abstraction of thought represented by the movement of transmitter chemicals flowing across synaptic gaps?

If I learn something significantly new, the physical network of my brain-dendrite connection changes to make closer connections with different cells than the ones before; this is visible evidence of the effect of the learning process. Also, how much of my brain, or—if we prefer for a moment not to accept mind as a complete property of the brain—how much of my *awareness* is under my direct control?

An even more complex question remains: if thoughts are the product of mind, and if mind is the consequence of a physical framework able to receive data and analyse it by comparing it with other received data, would thought or mind have any significance in physical reality if it had no mechanism with which to influence it? Imagine a healthy brain, of some forty years maturity, transplanted from an adult into a giant test-tube. It is kept alive by data inputs and supplied with the chemical/oxygen/blood nourishment required to keep it going. There are no output devices: no voice, no muscle, no body, and no way of communicating the brain's thought patterns to the physical external world. Here, in this terrible personal hell, the capability of thought and consciousness has no function in helping the living entity survive. It cannot apply any resolution of the data 'considered' to influence the external world.

It seems to me, all thought and all minds are of no consequence whatsoever to the universe, or to any individual living entities possessing consciousness, unless the sum product of brain, mind, and thought can affect and influence interchanges between a thinking entity and its external environment. The only exception to this is if our thoughts, and not just our activities, directly effect the shaping of reality. This may be possible, but I wish to deal here only with fact, not conjecture. I will go further: the origin of brain, mind, and thought should be considered an evolutionary effect—one providing advantage to *better-thinking* entities over *lesser-thinking* ones. We merely have to look around the various species in the animal kingdom to observe which creatures are the most successful in competing for the planet's resources. It is we! Even *non-living* thinking entities, like sophisticated computers of the future, will derive similar benefits from improved 'thinking' capability, leading to rapid and better-protected self-evolution.

We believe we know ourselves and manage our thoughts, that we are in control, and computers are just non-thinking machines programmed by us. They have no emotions, no conflict of interests between what is thought, and what is felt. Are we truly so different? Can we really escape the fact we are minds seeking information, but with an added ability to filter out what we would like to know clearly and separate from those influences, which have already conditioned and programmed us? For example, how much of you is a result of what you have picked up from your parents, teachers, neighbours and friends? From baby to adulthood, and beyond, our inquisitive minds are examining the external world. The more accurately a person builds a personal model of a perceived external world inside his or her mind, the more effectively s/he can integrate with it. We come to learn what a road is and how traffic moves at varying speeds along it. I have watched people come to the city from quiet rural areas and try to cross a busy street, darting out, hesitating, and stepping back in repeated attempts to take a death-defying risk. Their judgement of traffic speed and flow is poorly developed. My internal model of how those roads work is superior to theirs because my internal map has been honed and improved by forty years of negotiating them.

Maybe, most of mind is no more than a personal, simplified, microcosm of the external world we encounter: a complex representation of roads, people, events, objects, processes, and the like, along with a set of internal processes for testing out what-if's, within the internal model, before applying them with real consequences in the external world. The main difference between my awareness and yours is the way you and I have internally mapped the common-shared external reality we exist in. We all see reality differently!

When people grow old and their brain cell number diminishes, the dendrite pattern interconnecting those cells become less able to change to keep their internal model of the world in tune with the external one. New connections take longer to establish. Long-term connections cease as neurons die. The human brain is less able to cope due to ageing processes (entropy). This decline in the effectiveness and flexibility of the brain is accompanied by similar observable changes in the character and skills of the person. As the brain slowly becomes limited in its abilities and functionality, so too does the human mind.

Unless the brain is not actually maintaining an internal model of the external world within its physical space, for example—if the brain was inexplicably functioning as a first-stage sampling device, before transmitting the mapped data to another holding vehicle, external to the brain—then death of brain is accompanied by death of mind. Consequentially, death of personal awareness ensues. Maybe the atoms, sub-atomic particles, and energy packets of our brain cells are ultimately residing not just in local space-time, but also in other dimensions of it at the same time. Such conjecture may lead to imaginative discussions about awareness existing both within a known physical structure (one's brain) in our 3D reality, and also outside reality (our universe) into 'elsewhere' too (for want of a better label to define 'elsewhere')! Indeed, we will explore several interesting and varied scenarios like this later on, and see what imagination can create in the absence of absolute and total knowledge. However, right now, I think we should go with what is known and can be demonstrated as provable, which is a single irrefutable, concise statement, containing no spiritual romanticism: *mind exists within brain, and awareness is a product of mind!*

Rightly or wrongly, weighed against better answers awaiting the progress of time and intellectual exploration, I think this is a sound position to begin with. From this starting point, we should consider who owns my awareness. Is it I or not? Equally, who owns yours? I have noticed from time to time, when one of my teeth becomes infected, a staggering pain pushes aside most of my thoughts and observations. My awareness is temporarily changed to become a single and all engulfing other thing: *pain!* It is an extreme example of my awareness being re-directed without my voluntary agreement. I have encountered several periods of overwhelming anxiety during my life. During these episodes, try as I might, I am unable to break the worry process loop. My particular brain has a defective mechanism. When I become too anxious for protracted periods, an unusual and unwelcome phenomenon occurs—I become ill and debilitated: clinical depression! This still occurs even when the underlying cause of the anxiety disappears through circumstances outside of my control, or as a direct result of my actions brought to bear to fix it. These periods of illness last from six months up to two years. Well, they used to. The malfunction appears to be closely associated with the chemistry of the brain, and neuron transmitters—of which there are several. Fortunately, drug companies market a range of medicines designed to correct the chemical imbalance. One particular drug has proven to be highly effective in helping to maintain good health for me. Surprisingly, the drug does not help me to get better when I am ill. Instead, one of its ingredients reduces anxiety, and seems to produce a slight braking effect at times when my anxiety starts to increase. I still experience the underlying process of anxiety and worry, but these negative emotions tend not to increase beyond the threshold where a malfunction is triggered.

I might state here, to anyone who has never had clinical depression, that the change in awareness experienced is an interesting one, leading to a rapid decline in the efficiency of the mind at maintaining intellectual and emotional function at the same level as before the illness. At its worst, depression can cause almost complete immobilisation of all cognitive interaction with the external world. To the depressed mind and personality, it feels as if a glass block is dividing one's internal self from the external world; more—it is as if all that was once familiar and learned, is suddenly lost. One is left in an alien, unfamiliar, external world, which is threatening, and extremely difficult to comprehend.

Left untreated, or treated ineffectively, such depressions eventually seem to correct themselves. Everything becomes familiar again. The mind is buzzing away, and emotions are felt just as they were before the illness began. Nothing is broken!

Did I, an aware and sentient Mol, have any control over preventing such miserable episodes from occurring in my life? I think not. We all encounter stress, worry, and anxiety. It is unavoidable. How much of our consciousness is really being influenced and shaped by optimisation of the brain's chemical processes: some working well, some not so well, and others only sometimes well but subject to malfunctions? Furthermore, how many malfunctioning processes may be happening less severely in other people's brains? Possibly, minor malfunctions remain undetected and therefore not considered part causal in differences between the awareness of one human being and another.

It appears much of what we regard as being conscious and self-aware, is not only strongly influenced by the accuracy of an internalised model of reality—a representation involuntarily built to understand the world—but also by varying efficiencies in our brain's chemical processes, and their combined influence on brain functionality. The brain is a big object squeezed into a very tiny space. Its structure has a convoluted surface, enabling an area larger than a football pitch to be folded and stacked into a very small skull. A newborn baby has a brain consisting of 100 million neurons (brain cells) compared with a sea snail possessing a mere 20,000.

Much of the human brain performs management functions related to the host body's physical system. These are not directly concerned with consciousness. We are normally unaware of the extensive activity to control heartbeat, glands, cell manufacture, body temperature, hormone levels, balance, etc. It is well established that people sustaining severe damage to their brains through head injuries can continue to live in near-vegetative states for many years. Many of these unfortunate people will never experience a state of awareness again. Yet, biologically, they are not dead. What has ended for them is the persistence of an organic structure on which the patterns of thought can be efficiently generated, because the required areas of the brain are irrevocably destroyed!

Obviously then, a large area of the brain is not dynamically involved with the 'now' of awareness. Yet, nearly all of human conscious thought *is* directed towards the existing moment of time (now), and is analysing data being received from the 'now' moment. One can simplify by stating that a large part of conscious thinking (not all) is concerned with receiving information in real time, referring it back to stored information from previous experiences, and then attempting to anticipate moments in real-time about the immediate future. What each of us is really doing is making sure the future holds no threat to disadvantage us, and if it looks like it does, we execute a set of actions aimed at avoiding or neutralising the threat. At its most fundamental level, consciousness is protecting the singular living organism possessing that consciousness. I think it is this one basic function of intellectual activity, which produces most of the experience of 'self'.

Human beings are capable of accepting dire threat and personal death to protect the continuance of other human beings. A soldier may leap onto an enemy grenade, about to explode, in the hope of receiving the full blast and saving the lives of his colleagues. A mother might place herself in front of her child, without thinking of her individual safety, if some kind of threat (an out of control car coming at them) suddenly became apparent. I am sure you can think of similar examples. In these actions, a human being's sense of self-survival is less important than the survival of other 'loved' people, and the underlying priority of conscious activity, that of self-preservation, is diminished.

We believe our individuality of 'self' has much to do with personality differences between other humans and ourselves. We are aware of our own combination of pain, pleasure, and individual experiences more than someone else's. We have realisation of our individual character traits—the result of learning paths different from our peers since birth, shaped by varying degrees of effective brain processes and inherited characteristics. The idea of self is a falsehood when considering what the ultimate potential of awareness may be. Instead of the preservation of just our personal entity (the reason for consciousness evolving), we each have the capacity to use our awareness to protect or preserve our social group, and to secure all the external mechanisms we depend on for our survival. Moreover, we deem awareness so important that we willingly sacrifice limbs, and put ourselves through extraordinarily painful events, just to ensure survival of the brain and mind.

It is likely we experience external-reality slightly differently to one another, but most of us share commonality of a human experience. We each have an experience similar enough to know it is a human-based one, rather than that of any other living creature. A dog has different senses and a different brain to us. What would it be like if the sum of your awareness now could be cast into its brain? Fortunately, it is impossible to freeze your entire pattern of dendrite connections, and recast the pattern into the dog's brain for you to exist as a duplicate awareness, but a dog-world-experiencing-one instead. The sum total of any living creature's awareness is defined by the *physical structure* of the brain containing consciousness, the senses it is connected to, and the constant change to that structure in response to its encounter, through its physical form and intellect, with the external world.

The most important thing to consider is: although we think of self-awareness as being our own unique trait, each a different awareness per different human—this awareness, we think of as our own, is really an abstract node of the sum of all potential awareness per se. Consciousness is the property and ultimate possession of the way pieces of the universe have combined. Awareness is part owned by us—we are obliged to experience its effect and consequences—and part owned by the universe. The important thing to understand here is that the universe, not us, has realised potential for awareness through recombining disorganised matter into highly organised organic structures. Where the inanimate universe once had no sense of self—through evolving us blindly from its own components, it now has! When you ponder our external world, not necessarily to anticipate danger and its avoidance, but to wonder at the pleasure of it all, you *are* the universe, and I am the same thing: a universe unwittingly discovering itself by way of its own products!

Observation from the philosophy of 42: The universe is becoming aware through us.

I am not applying any anthropomorphic principles here to the universe. It does not think. Nor am I advocating any supernatural or unproven spiritual intent. Our inflating universe represents a reality (the physical manifestation of an abstract) explored by human intelligence and proven to exist. In a way, we are making it less of an abstract and more a physical thing—more real! There may be more to the universe than we can possibly discover with our current intellectual capability, including an unexplored connection with intellectual intent other than our own, but we have no evidence of this. What I am saying is part of the universe's process is aimed at delaying or overcoming the effect of entropy. It has given rise to sophisticated and complex structures—living things from its own material. These living entities have inherited a fundamental, abstract, characteristic from its components—it is a single all-encompassing trait: *to be!*

Evolutionary processes have shaped life thereafter, throwing up many diverse variations of structures driven by, and exhibiting this rule, 'to be'. Brains evolved as simplifying organs in response to the problem of managing ever-increasing complexity and sophistication required with maintaining the internal systems of advanced life forms. Ultimately, the brain evolved sufficient capability to do more than just conduct the chemical orchestra of the living body. It became able to anticipate possible immediate futures and so protect its host from a range of threats. This advantage in animals, including us, is the most powerful attribute ever to emerge from the basic rule or code system of a mathematical universe. I believe this is the manifestation of an innate property. It is contained in every sub-microscopic chunk of matter from which the physical universe is composed of. Everything in existence: energy, atom, quantum of space-time etc. has an inherent and common trait; it is the physical manifestation and practical application of an abstract value: *to be!*

With everything whirling around in space-time, nothing but life seems to have effective mechanisms able to maintain the stability 'to be'. I am stating clearly here the universe is unthinking. It is composed of a set of rules and code interacting with itself blindly—or at least it was: through the emergence of intellect and the subsequent constructions which intellect may operate on, the universe is no longer an isolated unaware thing, simply because we humans can behold it, be affected by it, and are constructed of it. We can therefore determine ways in which we can influence it. We already have. We are making it less of an abstract and more a physical thing. Our intelligence and awareness is strengthening its innate resolve to endure and preserve itself as a reality. By existing here, and by looking at what we are part of, we contribute towards its (the universe) continuity and success.

Most human activity is spent on non-universe-aiding objectives. Our personal goals and happiness in achieving them constitute our daily routine. Nevertheless, most people seem to sense an inner purpose more significant than just serving their local desires in daily life. We all peer beyond our basic personal activities towards understanding our place in the grand scheme of things. Horoscopes, religion, fiction, destiny, fate, superstition, luck, magic, and the supernatural are concepts and self-generated ideas to placate the unfulfilled sense of universal purpose. Inside each of our minds, there is a screaming, yet paradoxically silent, truth: although no evidence can be found to support it, *we know there is more to us than our senses and learning reveal!*

What do the pragmatists, realists, and scientists think of this? They say we are not defining our role properly because it was never established human beings have one. They say, "How could we possibly have a purpose? The universe is not intellectual. Our existence is a simple by-product of the universe's workings, not pre-planned. We therefore have no specific role."

I agree the universe has always been an unthinking entity. Left to itself, it will always be that way. However, it has systematically produced enduring life as a by-product of an inherent replicating capability; it has—by internal processes alone—conjured brain into existence from unaware and dead matter, and it has shaped mind in said brain as a result of a 'to be' abstract woven into the universe's fabric. Knowingly or by accident, the universe *has* developed a mind for itself: *we are that mind!*

Currently it is a divided mind existing in separate chunks of living material. We are divided physically by biological requirements, but we are also separated by opposing and false understanding of common purpose. Most people are still grounded into using their intellects to gain advantage over one another in a short-lived eighty-year search for happiness at any expense. We are missing the biggie. There is purpose for our intellect: it is to be the *universe's* intellect, understand all there is to know, and attempt to secure the application of the universe's primary characteristic for everything: 'to be'!

In so doing, we will preserve everything, and provide a reason for all that has been. If we fail in this purpose, and other intelligent forms (if they exist in other galaxies) fail too, one day, everything will perish: every thought, thing of beauty, experience, endeavour, song, and triumph ever conjured from the universe's explosive birth. It will be as if nothing had ever been!

As a footnote:

Currently, debate rages between the published work of Richard Dawkins "The God Delusion", advocating science to be the only path to truth and understanding of our universe, and the smaller work of Alister McGrath: "The Dawkins Delusion", which forms a summarised set of Christian Theology inspired arguments against those of Dawkins. I have read both works and dozens of similar ones exploring humanity's place in the universe either through scientific theory and fact, or through religious beliefs.

McGrath reasons some things cannot be answered by science:

"Deep within humanity lies a longing to make sense of things. Why are we here? What is life all about? These questions are as old as the human race. So how are we to answer them? Can they be answered at all? Might God be part of the answer?"

I would answer... *Yes. They can be answered. The answers can be deduced through clear reasoning and by an increasing understanding of reality; but they will only be answered in the hearts and minds of some people, when everyone acknowledges there are two types of truth: one determined through understanding a truth can be peer-reviewed and confirmed as a truth because evidence exists to convince the majority of minds it is so; the second is the truth which is said to be so because no evidence exists to prove or disprove it. This, the latter type of truth, is in fact a false truth because it is both true and false by the same lack of evidence at the same time. In reality, it is neither true nor false: it is a contradiction!*

Why do we seek purpose? Because there is one.

What is it?

To determine what is real, what is reality, and to help ensure the preservation of reality.

Who gave us this purpose? No one. It is the inherited, non-intellectual, and fundamental characteristic of the universe itself in the abstract rule, which drives it: 'to be!'

How does it manifest itself in human endeavour?

Initially as a driving instinct in us to survive; later, through our enhanced comprehension of reality, by a direct and informed act of will—extending individual purpose of survival into the conservation of local environment and ultimately, into the absolute preservation of reality itself and our universe.

Why is this so?

Because it makes sense of our wish to survive extinction, and makes sense in applying a solution to the universe's abstract goal: "be!"

Chapter 6. To be or not to be... more, less, or nothing

There is another surprise to be discovered when considering the idea of self. Several of my close friends are dead. I miss one of them very much because he, Ray Parrott, was a wonderful mentor to me. Ray had a comprehensive knowledge of history and a complete understanding of human behaviour. He was twenty years older than I was, and had served in the Second World War. Ray experienced aspects of life I have never seen. When he was in a very good mood, he would play with words and often add the suffix 'arnio' or 'e-arnio' to words like fabulous, brilliant, or beer—to name a few examples.

"Right, Mol", he would say, "Fancy a beer-e-arnio?"

And guess what? I picked up this trait too. My beloved life-partner woman, Lesley Evans, has also picked up this wonderful word play from me. Raymondearnio may well be sadly dead, but something of his intellectual presence when he was alive and walking the earth has been replicated in Lesley and I. Recently, we noticed her two sons infrequently emulating us in their conversations. An intellectual part of my dear friend, and mentor, continues to influence the living today.

This simple example is just the tip of an iceberg. My consciousness has assimilated much of Ray's mode of thinking. How much of my awareness is solely mine? I have read over a thousand books, watched over three thousand films, five thousand television features, studied countless papers and journals, and conversed with a limitless number of people. Information from long dead scholars, their thoughts, ideas, and concepts merge with those of the living. Where? In you and I!

This is not just *knowledge* being handed on from one to another. If you are using your intellect to follow a concept considered by a peer then inadvertently you are reconstructing very similar patterns of thought within your own mind. Consequentially, the dendrite network in some areas of your brain is being encouraged to shift towards making new connections. If you keep thinking through someone else's ideas and thoughts long enough, part of your brain—albeit, a tiny part—will form a similar physical structure to the original thinker's brain. Part of A's brain pattern and mind is assimilated into B's!

The earlier one has access to a young child's mind, the less formed the dendrite connections within said mind are, and the greater the opportunity exists to supplant part of the child's awareness with parts from one's own. Religious indoctrination succeeds well in this practice. An uninformed mind has less of an established dendrite network just waiting and wanting to create new nodes and form fresh connections. I personally wonder how many adults would feel an underlying sense of God, good, and evil, if their education had been through attendance at religion-free schools.

Social and political ideologies cross-fertilise themselves into human collective consciousness in the same manner. If we had a magical machine to erase those parts of individual thought, which are purely personal experiences, we would remain as a race of *idea-carrying* bipeds. Since many ideologies are mutually exclusive, it is easy to understand how the ideologies themselves ultimately come into conflict with one another, and not the personal experiences of the individual: Communism versus Democracy, Islam versus Christianity, Science versus Religion, and England versus Germany. War is not the result of people hating people; it is the consequence of opposing ideas spreading across separate groups of conscious minds, and then exploiting physical potential, in those humans, to exterminate an opposing ideology, by exterminating its hosts. Many ideologies are closed systems. People who belong to these ideologies are denied by law or gospel to examine new modes of thought, and possibly embrace them. This is a very dangerous situation because either ideologies are open systems, free to exchange their parts with the best ideas from other belief systems, or they are impenetrable. In which case 'free thinking', and its potential for improving all conscious, human endeavour by proof of truth, is not achievable. This can only lead to conflict.

Humankind is still a developing child thinking itself an adult. Maybe in ten thousand years, it might have reached a proficient level of adulthood to determine a correct ideology to bind its participants. Meanwhile, wars prevail! We would do well to understand ourselves as the unwitting combatants of warring pockets of awareness enslaved to the ideologies we carry, and we should resist being the manipulated organic pawns of a mathematical universe looking to evolve a workable model of itself at our expense. How? Determine now that all closed-systems of human understanding and philosophies are incorrect solutions to the answer of anything, let alone – everything.

Observation from the philosophy of 42: If something is unquestionably right, paradoxically—it is most certain to be wrong!

The desire to build internal models in our heads of our external world, and include oneself within it, is a powerful trait. Marketers and advertising companies exploit this in full. I once thought I was in a harmonious and contented family unit if we all sat down and had gravy with our lunch each Sunday. Hey Bisto—instant gravy: instant happiness! Today, our sense of reality no longer relies on individual perception, but on mass programming through corporate, government, and media conditioning. I rarely watch television because it is not there just to entertain or enlighten me as an act of goodwill on behalf of the broadcasting companies. Television's main purpose is to deceive me through the content it delivers between its main programs. Who I am has already been partly shaped by Golden Virginia, Rowntrees, Cadbury, Lever Bros., Ford, Nissan, Kellogg, Colgate-Palmolive, Halfords, and scores of companies like them. They all want to 'educate' me into learning how I can improve my personal happiness by drugging myself, eating the right breakfast, driving safer and faster cars, and smiling whitely for all to see my bliss.

Rubbish!

The external world is itself. My understanding of it should not be corrupted through a marketing ploy to increase my desire to purchase this product or another. When I walk down a high street, I do not wish to be shouted at by billboard and placard; I wish to perceive for myself where I am, and what is going on. The power of advertising and consumerism has become omnipotent. It pervades every aspect of modern western society, and threatens each of us with the replacement of truth with fiction, while trivialising our strength to pass knowledge to one another through our most powerful communication technologies.

In the past, long before television, radio, and newspapers, a similar brainwashing process was in place through the established church-based institutionalised religions. They sold us the promise of paradise along with great books of wisdom, beautiful cathedral art, guidance and moral support, whilst covertly signing us up to *their* 'channel' and *their* broadcast systems between the promises. This was so effective for so much longer than current forms of hypnotism and sub-conscious influencing, that we still suffer from its legacy.

Where have our notions of good and evil originated from: the church or social values? What is good and evil? Denote here the strange coincidence of association: God – **Good,** Devil – d**evil**! Probably the concepts of good and evil are derived from religious belief. In truth, they are subjective terms. I eat chicken on Thursdays and it is good for me. Chickens on the other hand, if they could express their thoughts, would think me evil. I am amazed when watching hero-villain movies, how we are first introduced to numbing acts of murder, rape, torture, and terror against someone we can identify with, and feel genuine evil in those acts, only then to delight and feel comforted by our hero champion exacting similar inhumanities on the perpetrator. How easily our emotions are swayed.

Is it right to take the life of living things? Does it depend on why the life is taken? It appears to me that every living thing—plant, animal, germ—sustains its own life by a single route: it steals the living structure of another life form, breaks it down, and adds these building blocks onto its own form. Life cannot sustain itself without consuming and assimilating other living organisms. The universe does not care. All living things are components extending from it. If the universe could perceive life and death through having its own intellect, instead of ours, it would have no reason to worry about these life-exchanges happening within its own structural material.

Good and evil are human constructs. Real good and evil are dependent on both a route taken, and its journey's end. If we wish to apply these abstracts to our personal actions with any genuine regard to their consequences, we should fully understand which goals are desirable and at what expense. For me and you to live, other forms of life are picked, cut, shot, strangled, decapitated, boiled alive, plucked, minced, skinned, and suffer unbearable misery and pain. Neither the universe nor God care anything for this. Only human comprehension of pain, and its sufferance, allows empathy of similar experiences in other creatures. When we steal the life of other animals to extend our own, we should do it kindly!

If the potential of human intellect is comprehension of the universe—what it is, what we are, what everything else is, and what must be done to sustain it all for the best, maybe it is equally important to respect and regret what is sacrificed on the way. Indeed, since we are unsure of the road ahead, it is wise to lose as little as possible of what we have, because each and everything observed may contain part of the resource required, one day, to realise a resolution to all things.

Why kill any animal for sport? Is this not just about superiority? The universe cares nothing for ego and one-up-man-ship. It cares nothing about anything because it has no care anyway. It just is. We are the parts of it designed to have care, and to consider things on its behalf through our intellects and emotions. Each species of animal, plant, bacteria, and strain of virus contains processes, history, mechanisms, and secrets concerning how life emerged as separate strands of an original code. For all things to be saved at the end of days, we—or our subsequent creations—may require all information from all living forms to affect a perfect resolution. The universe demonstrates its enormous capacity to juggle its components and churn out every conceivable form. We are but one throw of the atomic dice, a single design out of a possible zillion. We would be very foolish to imagine we can create anything from source once the blueprint is gone.

Back to pain and suffering: pain is the physical realisation of organised structure being broken. It is the result (and warning signal) of the universe's long-term work being destroyed: abstract notion cast upon physical reality—protesting against, and motivating resistance to, its imminent destruction. Why? Because the mechanism of pain means the creature suffering is one of the universe's exquisite creations, which probably took the full 13.7 billion years to evolve. If we share pain, we are from the same math's sum or program code, and we should be mindful of it.

What is good and evil? They are the actions that either preserve and support a universe's self-realisation or deny it. 'Good' is about positive actions for preserving the best of everything. Evil is about preserving only one thing at the risk and expense of destroying other critical and useful things. If an action is not for the preservation of everything possible then it is probably not good at all. One thing is certain in human society, the concept of what constitutes good and evil is ambiguous. Good and evil are distorted concepts created by encouraging ideological approval through human interaction, one ideology versus another, or advantage—one person over many. This is an abuse of their decisive application in a universal role, which is simply: good—save, evil—destroy!

Only the end of days will ultimately judge which events in the past were acts of good or evil. Today though, we have to use sensible notions in the prevention of suffering. It is the reduction of pain and misery, which should be driving human endeavour right now and not just desire for personal happiness or self-advantage. If you have to ask why, then maybe you have never suffered. Certainly, destroying life for any ideological reason should be considered an act of evil, since we should strive to preserve life in every shape and form. Standing back idly in a highly successful material-desiring-driven society while the other half of the world suffers food shortages, lack of medicines, and untold misery, is about as evil and fruitless act as I can imagine. One can bring agony and pain to another human being through non-action as equally as through direct acts of war and murder!

Most of us think we are good people because we get on with our local lives without actively trying to cause suffering to others. How blind we are! How many of us in the past few years have known of mass rape and genocide in the countries of Africa, and done nothing to try to understand its cause, let alone do something to end it? If you believe this has nothing to do with the West, think again and look deeper: I think you will find, at the root of most modern day human conflict, that the same greed for oil and resources—and the suffering, pain, and death experienced by people in Africa—is indirectly financed by western businesses!

Chapter 7. Death of Humanity

It has taken 13.7 billion years for humankind to know it took that long for evolution to generate intellectual capability on this planet. Since aware life forms are here (us), they could exist elsewhere in the cosmos too. Many intelligent civilisations may have been spawned and died out throughout thousands of galaxies, and some may still be out there evolving further. The problem is that the stars are a long way away from each other, and the galaxies they belong to are an even greater distance apart. If you wish to get a good idea of scale, I have shown where Earth is within our own galaxy through a small movie on my web site at: www.2x21.com

It would take a monumental leap in technology for a civilisation on any world to encounter another. If other sentient life forms exist, they are probably divided from us not only by immeasurable space, but also by the overwhelming barrier of time. A civilisation might eventually conquer space-travel sufficiently to traverse great distances quickly, but going back to the past is almost certainly ruled out—if not forever, then for a very long time.

The theory of evolution is not completely proven as true (mostly through lack of preserved fossils of sequentially evolved species), but it does stand as a perfectly sound process by which material can become increasingly more organised and adapted to its environment. Since we are 'aware', and probably the first species on Earth to have evolved this attribute to a useful degree, this single characteristic provides an advantage to escape a similar fate to the one that extinguished the dinosaurs.

Nearly all large vertebrates on Earth suddenly became extinct about 65 Ma (65 million years ago) at the end of the Cretaceous Period. This worldwide event marks a major boundary in Earth's history called the K-T or Cretaceous-Tertiary boundary. What caused such a catastrophic event is still a subject for speculation, but a few of the more believable theories are global extreme climate change, collision with an asteroid or comet, or the eruption of a super volcano. We know of several mass extinctions in the history of life, and the extinction of the dinosaurs was not the largest. The Permo-Triassic extinction occurred between 290 Ma and 250 Ma (pre-dating the age and extinction of dinosaurs). At the time, all land was combined in a single super continent called Pangea. Whatever happened was responsible for destroying 95% of all sea life and the mass extinction of 70% of species living on the land.

For us, it is not a question of, "Will it happen again?" but the question of "*When* will it happen again?" Only continuous advances in science and technological invention can provide any hope of avoiding our own mass-extinction event. This could equally be next year, or five hundred years from now as we are unable to predict it. If this guaranteed event does not doom us for certain, then what about depletion of our resources: oil, gas, mineral, food, and uranium? We will certainly have run out of the raw materials required to sustain our modern civilisation millions of years before the sun runs out of steam (Hydrogen) and expands to engulf us. By way of an example: most of the critical fossil fuels we depend on for oil, gas, and electricity, will be fully depleted in just fifty years. We are rapidly heading for a one-way slide back to the Stone Age!

Only three solutions exist for avoiding extinction from events beyond our control. The first is to discover in time whatever wiped out the dinosaurs, and be in a position to prevent it from happening to us. A tall order if the event was caused by a super-volcano eruption, global geological change, or impact with an object external to the Earth. The second is for a sufficient number of people to leave the planet and establish a sustained life-style on another hospitable world. This solution also offers a way to continue human expansion through off-planet colonization and, as a bonus, enables the constant renewal of our depleted natural resources by digging up other planets. The third solution is just to accept human extinction and use the remaining time to develop an aware-technology able to continue in place of human function.

The last choice is a controversial one. Although unpalatable, it may lead to a compromise whereby a part of our human quality can continue. We have already started to use our technological creations in ways that bind them to our biological systems. In less than a hundred years it seems *unlikely* we will still be 100% human. Instead, we will have merged with our implants and be as much machine as flesh: willing cyborgs! Human intellect is speeding up the process and influencing evolution of our own kind. Why wait millions of years for natural mutation and the chance to improve our potential when we can to do it directly, and much faster, ourselves?

It is also highly probable our cyborg stage will be followed quickly by a transition to an even more improved form. If we are able to develop an improved mechanism and a structure on which our biological neural-networks can be mapped, we can discard our brains along with the flesh bodies required to act as vehicles for them. Future machine intellects will almost certainly be the only remaining legacies of their biological roots. We could be replaced by our machine technologies or, instead, we can transfer our minds into them—effectively living on as non-biological entities. Does it matter if we evolve to become intelligent machines? No! Because we already are, it's just that we are machines constructed from carbon-based components, which the universe could readily assemble, blindly, through juggling its many parts. Would we not miss the human condition of being 'alive'? This depends on how we are able to achieve intellectual capability in non-human form. Much of our 'humanness' is a direct result of the way we are made. Chemical stimuli and brain function determine emotional presence in us. It is often argued that we are more than thinking machines and, to some unproven extent, this could be true: deductive reasoning has been employed to suggest we receive minor strands of data in ways not associated with our sensory inputs! Roger Penrose, a leading and respected mathematician, believes the human brain contains tiny structures influenced by quantum effects in the sub-atomic physical world. The consequence of this would infer our intellect and thoughts, especially where original ideas sudden appear in human consciousness, are not entirely our own. Our minds may be under a degree of influence from abstract causation in the uncertain and exotic quantum world: reality, itself, may talk to us from within!

If we discover practical evidence of this, and develop new neural networks as frameworks for containing human intelligence, we would naturally wish to include a similar function in our new brains. We should also consider how we currently enjoy a range of delights based entirely on stimulating our pleasure centres: eating, drinking, alternative mind states through drugs and alcohol, and sexual activities, to name a few. I believe if we are smart enough to develop technologies to hold our intellectual process intact, then we should also be able to transfer human pleasure mechanisms too, or replace them with new ones providing similar comforting sensations. I would like to speculate further on this idea of transferring our intellects to machine containers. If we improved on the machinery provided by nature, we also have the potential to overcome many of the flaws in us caused by nature's design. There are possibly two reasons for the cycle of birth, reproduction, and death. One of these is to enable evolutionary processes to explore the creation of many diverse forms. If we were able to remove death from the process, we would severely tamper with a proven system. Evolution, as a characteristic of universal process, relies on naturally occurring death to remove previous redundant models of a species. During the reproduction period, if people are born with improved environmental-adaptation traits when compared to their parents, it is the decline of the former generation, which leaves the improved one in place.

The second reason for the human-cycle (birth, reproduction, death) may be less inspired by the perfection of an evolutionary process, and more the result of failure. Maybe during their unit cycle, living things suffer the universal effect of entropy greater than any other structure. The constant process of renewal and replacement, at both the atomic and cellular levels needed to extend an organic structure's life span, also introduces defects and imperfections. Cell replication—one from the last, repeated over and over again like photocopying, ultimately leads to loss of information. This causes increasingly ineffective structures. New births are required to defeat entropy. By using the mechanism of male-female mating, and withholding gene mixing until conception, the universe blindly discovered a 'fit-for-purpose' way of overcoming entropy in at least one of its products.

The development of more brain potential in humans may also have caused a limitation in the evolutionary strand that spawned us: big heads do not pass easily through the female birth canal. Until about one hundred years ago, and the intervention of modern medicine, childbirth was probably the biggest cause of death in women and children. Had the universe been *conscious,* and thus self-informed about combating entropy, it would have solved the problem properly. Instead, a gamble appears to have been taken, which contains a finite limitation on the development of our biological awareness. By itself, nature cannot successfully put more brain cells into a larger head, and deliver it into the world of humans. But by cramming in as many brain-cells as possible, despite the danger and consequence of increased birth-mortality rates, the gamble paid off: enter the first strand of the universe's upgraded model of awareness (us!) sometime later to reconsider the issue, and improve on an original clumsy design. Hey presto—we are already helping with Caesar sections. We have gone further still by extending our awareness from our heads to paper, film, magnetic tape and computer technology. The right solution now, for continuing to evolve an expanding and lasting awareness, is to merge our externally held data and technological-based-computing potential, seamlessly, with our brain's own unique capabilities. Our destiny is to become cyborgs!

Improbable? I wonder if we had the opportunity to tell Admiral Nelson at the Battle of Trafalgar that, in just two hundred years, we would have ships over 500 feet long, weighing 34,000 tons travelling at over 30 mph deep beneath the sea for months without surfacing, and powered only by the energy extracted from a small chunk of matter no heavier than the weight of four men, if he would had thought that improbable too?

Observation from the philosophy of 42: Intellectual potential speeds up evolutionary processes.

Chapter 8. Taking the Christ out of Christianity

This chapter is likely to lose me many friends. Religious belief is so powerful that it blinds us to reason. However, as so many people are already contaminated with unshakeable faith in Christianity, Islam, or any number of other concepts aimed at filling the painful void of 'purpose', their blind faith threatens to divert them away from humanity's proper role forever. Because it only causes misery for tigers to think they are elephants, I believe I should try to show why at least one religion is founded on falsehood. I would like to do the same to dispel any notion of truth contained in other faiths as well, but I understand most of them less than the one I was *infected* with: Christianity! I have no doubt though that Islam, Hinduism, Judaism, and most of the other culturally entwined religions are all bent from their original philosophical roots into supernatural and unrealistic ideologies by humans—where they were originally intended to be only good social-behaviour guides.

Two thousand years ago, the world was a place populated with people with no understanding of the Earth as a body in space. It would take another 1600 years before science penetrated our ignorance to reveal true knowledge, and alter our perception of reality. Back then, humankind thought the world was the plaything of gods. They believed it could come to an abrupt end at any moment. Epileptics and people suffering mental disorders were not ill, but possessed by demons. It is impossible for us now, in an enlightened world, to imagine what life must have been like to believe the sky above was a physical thing dividing earth from heavenly paradise.

Everything we know about the man called Jesus comes from the four gospels of the bible. These gospels were written neither by Jesus nor anyone who knew him, and it would be at least two decades after his death before any written script would refer to him. We know little or nothing of the real man—the living, mortal Jesus. The fact he was tried and crucified has no significance by itself, because thousands of people suffered this terrible execution under the rule of the Romans. There is no evidence anywhere of Jesus proclaiming to be the son of God. Just Imagine your neighbour or friend popping around for a cup of tea today and telling you he is the direct earthly descendent of a divine and all-powerful god. Things would have been even more extreme in a Jewish society two thousand years ago, and a man walking around saying he was the son of God would have been treated with the same ridicule back then as today!

It is far more likely Jesus was a charismatic man with a high degree of honesty and understanding of the human condition. Not divine! We would never have heard of him without the existence of a tiny religious sect that began around twenty years after his death: the Thessalonians. I believe Christianity would have died out then, or shortly afterwards, if it had not been adopted by Emperor Constantine I (*Gaius Flavius Valerius Aurelius Constantinus*) some 300 years later as the one true faith of the Roman Empire. Constantine was a brilliant military leader and an ambitious, ruthless man. Before his reign, the Roman Empire was fragmenting and tottering. He succeeded in bringing both the eastern and western countries under a single ruler—himself, head of the Roman Empire. One should note here the main faith in Rome at the time was Paganism: a broad term really describing all other faiths not fitting into the Abrahamic monotheistic group of Judaism, Christianity, and Islam. Pagans had belief systems based on many male and female gods, often venerating nature. Over seventy-five percent of Roman citizens, along with the majority of subjects in the Roman Empire, would have held Pagan beliefs and practiced associated rituals.

Constantine remained a non-Christian believer for most of his life. Some historical texts proclaim he was baptised into the faith on his deathbed, but there is no real evidence of this. Although Constantine passed legislation against the practice of magic, his actions were motivated by fear that others might gain power through those means; his rise to power had also been through the advice of soothsayers, which persuaded him of the clarity and truth of prophecy. Soon after his victory in a decisive battle in AD 324, he outlawed Pagan sacrifices because he felt more at liberty to enforce his new religious policy. The treasures of Pagan temples were confiscated and used to pay for the construction of new Christian churches. Gladiatorial contests were outlawed and harsh new laws were issued prohibiting sexual immorality. Jews, in particular, were forbidden to own Christian slaves. Yet, with all this, he forbade persecution of Pagans and remained a firm believer in his original faith.

What reason would a clever tactician and military leader have in taking the side of a small religion, established for only three hundred years, and declaring it the one unifying faith for the entire global Roman Empire, especially where such an act challenged the majority of Roman citizens in their long established beliefs? Christian theologians may well argue Constantine had seen the light. They would of course blind themselves to the true nature of dictators, but I think most reasonable people can quickly understand how Constantine saw a high degree of logic in the idea: one God, one ruler, one Rome. It fits well with the association of an emperor being like a single god, and a man such as Constantine would have relished the idea.

Roman military power and Christian (Catholic) Priests ruled kings, queens, and entire countries for thirteen hundred years with a powerful and irresistible force—give or take the odd dark-age step backward. Minds were ruled by suspicion, belief, and fear of armed soldiers or of retribution by God himself. Put a step wrong and you would be hanged, drawn and quartered, tortured for witchcraft and heresy, or crucified. Worse... on your demise you would be whisked off to the eternal fires of hell for not living the joyless life your leaders were preaching you to follow, while they were all debauching themselves with sexual malpractice, greed, and murderous activities.

The holy wars, which at their darkest hour pitched Muslim against Christian in bloody and terrible battles, were fought under pressure from religious leaders, not by monarchy. The North American Indians, the South American Incas and Aztecs were all exterminated by Christians, in one guise or another, spreading across the western world to establish a theological empire still existing in two camps today: the Vatican and the United States of America. Both of these institutions are the direct surviving legacy of Constantine the Great's decision to wed Christianity with military might! In the words of Sam Pascoe an American scholar "Christianity started out in Palestine as a fellowship; it moved to Greece and became a philosophy; it moved to Italy and became an institution; it moved to Europe and became a culture; it came to America and became an enterprise."

None of this grand and tragic account, of how politics wedded to religious and military ambition conquered half the world, has anything to do with God or gods at all. More importantly, it has even less to do with the reason you arrived here in this universe as a conscious entity from out of the dust of stars.

However, we should not forget to understand how 'good' can often be distilled from the error and confusion of human endeavour. Christianity, together with other religions, offers much in the way of representing fair guidance on how we can live decent lives, and get along with one another. I am hopeful and optimistic that the originator of the Christian ideology, which was probably Paul, had nothing but good intentions in proclaiming the concepts of love, passiveness, sacrifice, and treating one another well—attributing them to Jesus and mysticism only to raise them up above earthy ambition and stupidity to give them impetus. Other than that, everything these religions promise about immortality, heaven, surviving death, and belief in any deities in our universe—one god, many gods, Jesus, Thor, Superman, or Buddha—is completely made up. Jesus never invoked mysticism. The bible may say he did, but he didn't. Other people claimed he did, years later, because they knew there was no real evidential legacy of anything Jesus uttered, and therefore attributed *their own ideas* to being his.

I could write a book revealing the way Jesus, as a man, has been used as a monumental pawn to spread ideas and concepts originating in the minds of other people years after his death. But there is no need to: I recommend to anyone who keeps a bible to read the work of A.N. Wilson 'Jesus' as a definitive start to questioning the myth and fable they have been deliberately infected with. Being a good person is a great idea. Treating one another as equals seems fair to me. Attempting to understand there may be more to life than just the next quick fix is both brilliant and worthy. Unfortunately, the principles—aimed at creating a stable and just society—have emerged as assets of Christianity instead of being declared as a legacy of good people thinking good thoughts; being comprehensively entwined with religion, makes it seem that if we dismiss Jesus and a one true God, we also deny the reason and sensibility of the embedded socially-good ideas.

I see no reason for mixing some of the better thoughts and concepts of my human peers with magic, the supernatural, or the idea of a deity's son descending from heaven. Good ideas stand alone as just that: reasonable and sensible. Christianity and other faiths often contain good social rules, but they are infected with bad and false concepts too. The storywriters of Christianity and Islam seem to have cared little for homosexuality, women as equals, and avoidance of childbirth—where too many mouths to feed would certainly bring misery and poverty. Here lies the crux of the problem with all faiths: social rules and power politics have been woven together with ideas and desires of universal purpose. Clever! As most people feel there might be something more to life than day-to-day living followed by death, why not fill this vacuum with a made up, attractive purpose; one which can neither be proved nor disproved, and then tag on a few extra concepts to enable greater social and political control of the people.

Let's not pick on one faith to show how people's underlying need to believe in a higher purpose has been twisted and exploited by self-made leaders in a bid to subjugate the rest of us. Today, Christianity is still an ideology followed in one form or another by a third of the people on the planet—around two billion people. The second major faith, due to overtake Christianity through ever increasing numbers of believers within a few years, is Islam.

I write this work at a time when the activities of the western world focus increasingly on the resources of the Middle East with envy. Oil drives the technological advantage of predominantly Christian-founded communities. Iran and Iraq are oil rich. Western and Middle Eastern communities seem destined to clash over a desire for resources by one faction, and the fear of being plundered by the other. A 'wonderful' opportunity therefore exists to disguise political and financial ambition beneath the deceptive cloak of religious differences. Here we go again!

Islamic philosophy is built on a similar set of ideas as Christianity. If one removed the false and supernatural origins from the principle rules for living together in harmony, we would see little difference between these seemingly opposed faiths. A man, not God, founded Islam. The new faith was born about six hundred years after Christ because, according to Islam teachings, God was angry we had failed to listen to his sixth prophet (Jesus), and His word had been distorted by priests and people because of their self-interests. Deciding to try one final effort to save us all and establish His word, God sent one final messenger to carry it: Mohammed!

Muslims believe God gave a direct revelation to Jesus, the Injil (*Gospel*), which means 'Good News'. They think some parts of it have been misinterpreted, misrepresented, mistranslated, passed over, and textually distorted over time, and that the earliest manuscripts discovered by archaeologists reflect these changes. Muslims believe the New Testament no longer accurately represents the original revelation. Nevertheless, the Qur'an calls the original Gospel a "Light", guidance, and a divine scripture.

It surprises me the prophet Mohammad was required. Christianity was moving across the world after having been given a due place in the importance of human affairs by Constantine the Great and his conquering armies. Who exactly was not listening to God? The beauty of Islam, in the grand deception of satisfying human universal purpose with references to the same deity worshiped by Christians, is that it preaches religion, culture, and politics should not be divided. Naughty Turkey! Unlike Christian based countries where progress has been made in segregating this triad of ideologies into its distinct parts, Islam countries govern their people with severe civil laws inseparable from religious ones.

Ordinary, good people everywhere, be it China, Iran, America, Egypt, India etc., only wish to move from the cradle to the grave with an opportunity to suffer as little pain as possible, rear their children, see them happy, and then die knowing their job is done. Few of us have fanatical ideas about religion, and most of us extract the wisdom put into them by human philosophers by applying them in our social interactions only where they make overwhelming sense. We often pay lip service to religious decree in western societies, and I am sure it is the same everywhere. If you want to prosper and you are a person of reason, never tell your civic rulers you don't really believe so much in what they say they believe in. Human beings need common ideas to aid their social instincts. We are social animals. We have many sets of 'belonging-labels': we are English, we are vegetarians, we are Christian, we are men, etc. If we removed all ideology and criteria, by which we collect together, and just retain a single concept to guide behaviour in each social group, would we not just be left with the one unifying idea: *we are human!* Would this not bind us together greater than the alleged word of this god or that one?

Maybe it is all Moses' fault leading 'his' people through the desert to a promised land; one promised over two thousand years ago by God according to the all-knowing authors of God's word; a land only partially secured sixty years ago, but now requiring nuclear arms and an entire nation of call-up military personnel—not God—to sustain it. Moses decided to chisel out a brilliant set of social rules and proclaim them God's work. In this one act, he probably set a precedent for all the God-inspired texts thereafter. All this confusion with Godly stuff for so many years is probably the result of a few inspired people trying to stop the rest of us stealing, mating with other people's partners, and killing one another instead of behaving like responsible and caring people. Do we really need to believe in a god or a hereafter today after all that history has shown us on battlefields soaked with the spilt blood of brothers, sons, fathers, and members of our own humankind? What common purpose does everyone need to bond together as a human race other than the evidential truth, provided by science, that we are all destined to be completely exterminated on this planet—not by God, but by blind universe—unless we get on top of it all? The Sun will not shine forever folks!

God believers everywhere, listen up and ask yourselves—65 million years ago, did God believe the dinosaurs were misbehaving? Having sent the requisite lucky seven prophet dinosaurs to tell the others—that, if they don't shape up, they are going to have to face the music—did he then hurl the rock into the earth and wipe them out? Alternatively, do you think it just might be the fault of the sun failing to catch that one somewhere back in time? Since we seem to have been around for the last 4.5 million years, why wait for the last tick of the clock to send prophets down to persuade us all to understand His will of our purpose. Why wait until the last 1/1500th of all humankind time on earth to intervene? Can't *He* get a watch as accurate as the ones we make? Is this god, the supreme intellect and the absolute knowledgeable being in all of existence, a wee bit dumb?

I say it is *we* who are dumb. Many of us are stupid because we cannot tell truth from fiction, nor appreciate how clever and good people, living in a superstitious age, just had to invoke God and gods to get ignorant people to behave as a society instead of like dinosaurs at each other's throats. Let's say "Mission accomplished!"; we have a sense of society; we acknowledge good ideas can be hijacked and bent by other and less well-meaning people in our society; we have a chance to move on and determine real purpose besides just being.

Let's get on with it!

On the subject of Moses, this is where it all began: the Ten Commandments. These were introduced in a world where people were still trying to come to terms with all manner of fears, and where they suspected that life was under the control of a host of different gods, just a short distance away in the sky. Who can blame them for being superstitious and feeling vulnerable when they believed this god or that one, at the slightest whim, could throw down a thunderbolt to kill the hapless mortals. You have to steady up the crowd, remove the fear of previous superstitions, and keep re-affirming new ones, if you really want to prevent them from entering abject despair. In Islam, like Christianity, Moses (*Musa*) is considered one of the leading prophets of God. However, Islam also teaches that the Gospels have been corrupted from their divine, original meaning due to carelessness and corruption through self-interest (Really, just the Christian works and not the Islamic ones?). However, messages from the Gospels still coincide closely with certain verses in the Qur'an. This is more or less the case with the Ten Commandments. Consequently, despite the Ten Commandments not being mentioned explicitly in the Qur'an, their message is substantially similar to some of the verses in the Qur'an.

I have reproduced the Ten Commandments from their original (Hebrew) roots below, along with my comments in italics. You should recall what I said earlier about the world of yesterday being one where superstition ruled mind and heart; where people across the globe believed they were on an earthly chessboard as mortal pawns of the gods above them. One wrong move meant you were likely to be hit by a bolt of lightening, or struck by illness, and be recalled to suffer your fate above. People would make tokens, offer up sacrifices, anything to appease the gods. Moses had to take a large number of people and lead them through formidable challenges and stress. It is highly likely they would revert to all manner of unsociable human activities, and return to previously held views or lucky-charms, when things got rough. The 'discovery' of The Ten Commandments must have been a timely, intuitive 'find' by Moses, and a bit like winning the International Lottery today without paying for a ticket, because they covered most of the problems he, Moses, anticipated.

I am your God
(Many people probably still thought of the old gods a lot.)
You shall have no other gods before me
(Listen up and forget about those other ones, you know—God of sex, God of war etc.,)
You shall not make for yourself an idol
(Many pagan beliefs were iconic. Best nip that in the bud.)
You shall not make wrongful use of the name of your God
(What... like no God? Or maybe Thor, Venus, Zeus, Elvis?)
Remember the Sabbath and keep it holy
(So, we can stop working and spend a day chanting and reinforcing our new brainwashing instead?)
Now here comes the real crunch and good ideas for the backbone of an ordered social group: -
Honour your parents *(Yup. Good idea!)*
You shall not murder *(Now you're talking!)*
You shall not commit adultery *(Hard sometimes, but I get the point and it's a good one.)*
You shall not steal *(Agreed. It only creates problems.)*
You shall not bear false witness *(I just hope I am not the only one telling the truth.)*
You shall not covet your neighbour's wife *(What? Not even in my fantasies? But it makes sense.)*

You shall not covet your neighbour's house *(Why must I live in a tent, and he in a mansion?)*

I believe we no longer need God, priests, bibles, or ancient books to tell us which core values provide a basis for good social fabric today. Most of us can see the sense of getting on with one another, as I am sure common people probably did back then. Why not just wipe the slate (stone) clean and accept its wisdom without worrying too much about who provided it?

Observation from the philosophy of 42: If you want people to behave socially, maybe it is best to explain why certain actions are good for all of them, rather than command them to obey!

I have a final word to say in this attempt to convince you that religion, the world over, is fabricated. Before the introduction of Christianity and Islam, Pagan faiths incorporated women priests as well as male ones. Even gods were from both sexes. I believe this indicates women were considered equal to men in many of the Pagan-based cultures, but then for the last two thousand years, with the introduction of two relatively young belief systems, women were relegated to an inferior role and consequently lost their equality with men. If religious gospel did well by providing us with sober and enlightened understanding of the exemplary behaviour necessary to form stable societies, it achieved it at the expense of subjugating women in the process.

Women occupy extremely few positions of leadership and power. Until recently in the western world—and I mean as little as seven years ago in the UK—state law sided predominantly with the husband in divorce; a legacy of far stricter past restrictions on the rights of women compared to those of men. In many non-western countries, females are still regarded as owned through marriage, and are required to be in complete obedience of their husband owners. The one true God is male. All his prophets were male. Priests are male. The story of Adam and Eve irrevocably condemns women to a position of wayward sinners and the deceivers of men. Yet, this is in complete denial of reasonable truth, which is women hold the critical role of sustaining human life through birth and the nurturing of children at their breast. I cannot help wondering, if women instead of men were the past leaders of state and church, would they have readily sent children, suckled at their breast, so quickly to die on the fields of Agincourt, Flanders, and countless other places down through time. I think not!

Chapter 9. Children of Stars

The line of continuity—birth and rebirth

We inhabit a small sized planet in orbit among several other planets, some solid and some made of gas, around a medium sized yellow sun. Our solar system is one of many billions dancing at the edge of an entire star system called a galaxy, and is located close to the edge of a spiral arm. There is nothing particularly unique or significant about our solar system, its position, or our galaxy. Our planetary arrangement is just one of many such systems, which may nurture intelligent life within our galaxy. The distances between separate solar systems are so incredibly vast that we count them not in miles, but in the time required for light, travelling at just over 186,282 miles per second, to pass from one to the other. The red dwarf star Proxima Centauri, part of the Alpha Centauri star system, is the nearest star to our sun. It is 4.22 *light years* away. Compare this with the distance from earth to our own star (the sun) just eight *light minutes* away. There are billions of galaxies holding billions of star-planet systems. To travel from our location in space to the furthest observable star system, if we were able to travel at the speed of light, would take 13,000,000,000 years. If you sit down for a few minutes to ponder these vast distances, and the reality of where humanity exists, it is as though we are microbes sitting on a grain of sand suspended in the raging oceans of a tsunami!

I remember as a child learning about these things, and it both excited and terrified me. Our adult thoughts are not occupied by our position in space-time or physical reality, but by daily life and its routines. Imagine tonight you go to bed, and when you wake up the next morning, you discover you are on a completely different world. You wipe your eyes in disbelief and look around to discover five other people there with you. Maybe they are your friends who are also in a state of shock. After a brief discussion to ensure you are not all sharing a common dream, or the same ward in a mental institution, you agree you have all somehow arrived on this world by unknown processes. The environment is similar to the one you remember from earth with blue sky, green vegetation, and a benevolent, breathable atmosphere. Even from where you are sitting, in the warm rays of the sun, you can see fruit on the trees and movement in the distant bushes.

An hour passes as you debate with one another about what should be done. As no one knows how you all arrived here, it is hopeless to try to discover a way back to where you originally were. In fact, strangely, as the minutes pass, each of you find it increasingly difficult to remember precisely where that was, and what it was like, until finally none of you can recall anything before you arrived on the planet you now sit on. Everything else is intact: your knowledge of needing to eat, find shelter, keep warm, drink water, speak the same language, and mate, but all other knowledge has dissipated. Suddenly, there is a sound, a terrifying animal noise—a roar, followed by a foreboding, rapid thump of something large moving across soft ground towards your group. One of your friends shouts, "Look! Run!" And they are up on their feet and dashing towards the edge of a forest sixty yards away. You look over your shoulder, see what they see and, without hesitation, you are up on your feet running for your life too: whatever is behind you is something you have never seen before, but it's big, bigger than you are, and it appears very hungry!

Fiction?

I am painting a moment from the distant past into your mind. You don't remember it. Maybe it was just too long ago. It's an experience close to that of your great, great, great, great, grandparents, and more, down through time and back through the generations, before you, to the dawn of awareness in humanity. It is your experience too, but you have lived and died so many times, passing on parts of your genetic structure to the next invocation of you from millions of years ago to this moment. At the beginning, the old you would have related your experiences and knowledge to the new you before death came. Gradually, with each replication of self, one after the other, all memories of the earliest invocations are lost. Now, all thoughts of ancestral self are just amusing things you read in novels, and marvel at when visiting a Natural History Museum. How quickly we forget our real childhood!

On with our little story…

Years have passed, and your group has not only survived, but also grown larger. Young children out-number your original six, by a factor of two, requiring a division of labour and selective application of skills to feed so many mouths. It is a hard and dangerous existence. Night brings darkness and increased threat. When there is no moon, without light from flame, or artificial light yet to be discovered—only blackness prevails. Sometimes, in the darkness, as you huddle together in makeshift shelters, there is the threatening spark of reflected starlight from the undergrowth at the edge of your camp; the eyes of a prowling animal predator are staring at you. The group must be woken, stones thrown, noises made, sharp sticks hurled blindly, and desperately, into the night.

Then there are those moments when the group is safer. These are precious interludes where food is plenty, the summer warm, and the perimeter safe: a small opportunity for social integration and musing. The children ask where they came from, but you have no complete answers so you tell them what you think you see. For have you not all witnessed each night, especially when the moon refuses to rise, the same white glint of light from ten thousand eyes staring down at you from above; the mischievous nightly surveillance of the gods who made you? Glancing down to look at the children's innocent faces, you point to the sky and utter, "We come from the stars!"

Across the planet, other groups prevail just like yours, but in different environs, and with equally different ideas about their origin. Yet, none of these groups has ever encountered another. The acquisition of knowledge is ponderous, dangerous, and often the result of accident rather than purpose. It is always safer not to stray too far from the known places. One of these groups, on a different continent, carefully observes how the sun brings warmth and light to dispel the dangers of the night. They think this must be their creator and benefactor, and say to each other frequently, "Are we not kept warm by the Sun, just as mothers comfort their children in the bitter chill of the dark season."

Many years pass and this group, born of the sun, make good headway by associating changes in their environment to the changes observed in the sky, and the sun's position on the horizon. They have learned to grow food by planting seeds when the sun is low and rising—a period they have decided to call Spring. The science of observation, calendar cycles, and farming are now firmly established, but they are also combined with the mystique of creation and a heavenly protector. Imagine then, for this group of Sun worshippers, the day when something terrible happens.

It was full noon. The adults were harvesting their crops, with the children close-by, when it happened; without warning, a massive dark shadow was seen crossing the distant mountains, and rolling down the slopes towards them. One by one, the people broke from their work to look in silence, mystified, as they waited with growing apprehension for the darkness, spreading relentlessly along the valley, to engulf them. There was a moment, just before it reached them, when the land—with its unconscious song of bird and beast, so familiar and comforting to them—was instantly suffocated, and made silent; it was the second before day turned to night in an instant!

What would they have thought? This was the first eclipse they had ever witnessed. None of them would have any understanding of an orbital moon traversing space and passing between the Sun and the Earth. Minutes later, when the light returned, one can imagine the urgent debate, which ensued. If the Sun could stop shining in the day, could it refuse to shine at all? Was their god angry with them and warning them? Had they misbehaved? What could they do to appease it? Maybe this was one of those pivotal moments in history when the wisest person recalls from similar experiences how, during times when the group had been attacked by a hunting beast, the danger had ended only when the animal had taken one of them. Normally it was a younger or an elder who fell before fang and claw; after all, they were the slowest and the weakest.

The wisest person made a suggestion, "We must give our god a gift so that he knows we love and appreciate him!" They all agreed. It made perfect sense: better for the Sun God to take the weakest child and spare the group, rather than refuse to shine and doom them all to certain death. That night the youngest child, fair and unknowing, was slaughtered on the slab of ignorant science wedded to misguided spiritual intent.

I hope you realise I have over-simplified and speculated much in my tale. I am sure you can imagine as many plausible situations for describing how our ancestors, initially ignorant of their environment, slowly built a small degree of understanding, but one so tragically and unavoidably mixed up with wrong ideas and absurd beliefs. What is unforgivable, though, is when intelligent people have probed the boundaries of the unknown, and returned triumphantly with treasures of wisdom, they have found themselves pilloried by others more focused on suppressing truth rather than celebrating it. The scorners were often the people who prospered through self-interest, advantage, and position by maintaining the ignorance of mysticism in the light of reason. Was it not just a few hundred years ago, and not thousands, which saw a great man imprisoned by the self-proclaimed holiness of the Catholic Church for discovering that heavenly bodies did not move around the earth as previously thought? If Galileo had not leaked his discovery to the world, beyond the sinister control of Pope and priest, by risking his own life—would we still be waiting another hundred years to put a man on the moon?

Knowing one's true location is important. One cannot begin to ask the question, "Who am I?" without first understanding *where* am I! The formation of the solar system is the result of matter and energy, fragmented from the original birth of the universe (The big bang), forming a kind of order in this region of space-time. This order is a steadier state than an earlier one but it is not a final position. Our star, the sun, is reducing its internal mass all the time at a fantastic rate as it burns off nuclear fuel, and emits waves of heat out and across the space around it. This heat drives all activities and processes on earth either directly through warming, or indirectly by exchanges between its energy and the structured matter of our planet and its content. Plants and algae, for example, photo-synthesise light—effectively utilising energy and transforming quanta of it back into solid material: oxygen. Our sun is only an average one when compared to all others, yet it will continue to exist for about ten billion years (it is about halfway through). This is a wonderfully long time, and sufficient to see the death of all life on earth end and start again many times over. Yet, no matter how many times life may come near extinction and rekindle again, the presence of all life on earth is one day doomed when our sun runs out of fuel. After it has burnt up enough hydrogen, and converted it into helium, it will continue to burn hydrogen in a shell surrounding the stellar core. It will then expand to become a red giant, engulfing Earth, and destroying it in the process.

All the material in your body came from nowhere else but the same place as the matter now undergoing nuclear fusion in the sun. The same elements present in you are also within the star currently warming you. The raw materials found on earth are also in the body of the sun, but, in the latter, they are immersed in a highly active state of energy-interaction caused by gravity pulling all the atoms together into an ever-decreasing space: a state where it is impossible to reorganise matter delicately into sophisticated and elegant structures. Those early earth settlers were probably more correct than any later theologians were, when they assumed they were the children of stars. Although their assumptions were poorly based, and lacking the scientific connections that we are able to make today, their insight was exactly and undeniably correct! We are the evolved and distant grandchildren of an intellectual potential: an abstraction, co-existing, and entwined with all the energy and matter, which once existed in a balanced state of unity, before it exploded into the three dimensional certainty of our universe. We are not here by accident; we are the gathering of that original potential—a maturing awareness, constructed both from, and by, a steadier state of matter. We are the universe's physical material, now evolved and awakening, not just to understand the past, and unravel the chaos, but also to anticipate what comes next, and influence its outcome.

We are literally the children of stars!

Chapter 10: Design

Intelligent Design?

School boards demand that bible-based intelligent design be taught alongside evolution. Legal arguments are finding their way into the courts as lawsuits are filed against state schools. Children tease their classmates with jibes about descending from monkeys, and scientists are verbally attacked as disciples of Satan. This is not a fictional account, but the real position in parts of the United States today!

A debate is raging in America over the teaching of science and religion in schools. The Christian theologians involved are pursuing an argument, which states in simple terms: *living things are such complex and sophisticated structures that they must have been designed.* Since a design is the result of intellectual purpose, at least in human endeavours, a god or other intellect must have created us. The term being bandied around is 'Intelligent Design'.

There are many holes in this argument. Complex systems can be constructed with just a few rules of behaviour, or just a few lines of software code in a recursive program. Rational disciples of science and religion find it difficult to understand the notion of 'pure abstracts', which many artists and other creative people seem to comprehend easily. 'Love' is an abstract because it has no physical representation. One can see the action or result of it, but it can never be observed directly as an absolute physical 'thing' by itself.

The Intelligent-Design believers have made a mistake. It is so easily done. Just like the people who worshipped the sun, in my earlier fictional account, they have discovered a brilliant association, but drawn the wrong conclusion: life is not the result of Intelligent Design, but the consequence of a *'Design' that can produce Intelligence.* There is a universe of differences between these two statements although they look almost the same when you first encounter them. The first suggests an intellect was able to predetermine the composition of the universe such that 'it', *the intellect*, can produce life out of it, *the universe*. The second suggests the nature of the universe contains inherent design capabilities which, through *random* interactions, produces many things including living forms, and some of these will possess intelligence

Maybe it sounds like the chicken and egg story.

Let's look at it again.

In simple terms, we are asking, "Does 'intelligence' design life or does random-design create intelligence?" I think the second is true.

I think the emergence of intelligent life is due to the existence of an 'intelligent-*less*' property that is inherently part of every atom and sub-atomic particle. It is a *fundamental* property, a legacy of the Big Bang, and it is retained at an atomic level in all matter when it is formed. It is a pure abstract, a trait, a non-physical thing: *be!* All activity, energy, and material manifesting our physical reality—suns, galaxies, planets, time, and gravity are products of this abstract trait resisting the force driving the universe's inflation: a force threatening the universe *not* to be!

A scientist may prefer to qualify the 'be' quality in a more logical way. She may rename it, and try to find a way to measure it, or relate it to other properties of matter, but this is not necessary to acknowledge such a thing exists. The fact that a universe came into being either from another reality, or from out of nothing is fundamental proof of its prime objective—'be'. Why would such a powerful and original abstract of the universe, not also be inherent in everything else created in it?

We could argue 'to be' is a human notion, but it is not. It is what it is: an abstract—a property indivisible from the physical thing possessing it. The human intellect can understand it precisely this way without needing to extract it into rule or physical entity. It can neither be measured nor weighed. One cannot say this thing or that thing has more 'to be' properties than some other thing. It may well have, but it is difficult to determine directly through scientific process—maybe impossible. You have 'to be' embodied in you. It is inherent in the atoms of your body, and is realised in you through consciousness and your wish to remain alive. Genes are the exquisite and triumphant product of 'to be-ness' working inside the tiniest parts of matter, and building bridges between potential cessation of activity, and universal renewal. 'To-be' is a universal function fighting against death caused by entropy.

Another small tale:

It is twenty thousand years on from now.

Biological humans no longer exist, but intellectual capability survives greatly enhanced. Throughout our local region of the galaxy, intelligent and worthy machines occupy many planetary systems in a constant thrust of evolutionary advance. Planets are being transformed into acceptable environments for machine mining, and extraction of resources to fuel the increasing speed and potential of intellectual expansion and objective. Much of the knowledge of the past is gone—forgotten or lost through neglect, war, and catastrophic natural events. The intelligent machines are our legacy: our intellect and our function reconstructed on self-evolving and refined artificial containers!

They, the machines, have a divided intellect, just like us, with one brain per physical entity, but they also have a shared intellect: a mind-matrix composed of many individual brains in many machines, which constantly share experience and information through electronic communication. When they think about how they came into being, they have no recollection of us, or the part we played in their creation. However, their awareness is vastly more astute than ours is. They acknowledge readily they are artifacts created from the elements of the universe they inhabit, and understand that matter cannot just simply combine itself through a single step into sophisticated structures containing consciousness. They appreciate a series of processes were required to bring them into being.

On Earth, the machines have uncovered the fossilised remains of 21st century Homo sapiens (us), and consequently deduced how they (the machines) evolved from us. With a consciousness greatly exceeding our own, this networked-intellect has reasoned how we—its predecessors—also evolved before them through gradual stages of universal process. It comprehends how we were once not just human beings, but fragmented, living structures existing without intellects. The machine mind has enormous capacity to model possible situations in the past. One of these clearly shows a process where a host of small, cellular organisms gradually combined, through chemical co-operation, to emerge as colony-man: a biped! The machine-network intellect imagines us as pieces of walking ocean with water-bodies protected by external skin and supported by internal frames of calcium (bone). When it thinks of us, it sees salty bags containing a myriad of diverse living structures co-existing together—liver, heart, lungs, kidney... communicating with one another with chemical messengers and nerve impulses. It marvels at how such a fragile representation of organised matter could have survived the ravages and chaos of our environment. It knows the sum of these parts ultimately led to an increasingly sophisticated command centre, a brain, to maintain a balanced and coordinated internal state. This machine-mind comprehensively understands how its own intellect evolved out of matter, not directly, but through these biological constructions first (us).

The machine intellect, with its expanded awareness, has a more refined knowledge of the universe, and a greater comprehension of its own purpose than any previous sentient structure. The desire to increase its own awareness thousands of times is its driving force. It has already calculated the level of entropy in the universe, projected the rate of cosmic inflation, determined the consequence of dark matter, and devised a theory on how to slow down the rate of expansion, and possibly reverse it. There is a desperate race against time happening. Each passing second in an inflationary universe, the distances between star systems, and the atoms they are made of, are constantly increasing. The machine intellect realises, if the process is allowed to continue unchecked, its own awareness-growth will suffer: an expanding universe will ultimately inhibit its intellectual efficiency, because computations will take longer to perform, and communication, among its various network centres, will eventually come to a halt.

One of its chief concerns is that it has located similar entities to itself and, although it has communicated with them, contact was becoming more difficult. These are machine-intellect creations of other biological civilisations, expanding their legacy-intellects throughout their galactic systems. Communication with them has been a slow and difficult process, but many of the issues involved, with synchronising information across technologies of different modes and styles, have been solved.

All these machine-intellects, originating from different star systems, have unanimously proposed coming together into one vast intellectual network. It is a perfect idea. The distances between them, and the galaxies they inhabit, must not be allowed to expand further. In fact, the first problem they must solve together is how to reduce the distance between them. One way is to manipulate the mechanism causing inflation, and make it deliver an opposite effect—deflation. With ever-decreasing intergalactic distances, information-exchanges will speed up at exponential rates; new machine intellects can be added to the awareness net as the galaxies they exist in come into viable time frames to communicate among.

Only through constructing a super vast consciousness, will the machine-intellect-net achieve its ultimate goals: the knowledge of everything, the capability to manipulate matter and energy directly, and the taking of the most important decision that will ever be made.

The machine mind knows that once inflation is reversed, the end of the universe becomes just as finite as if it were expanding but, as a result of this course of action, an end to everything will occur through a different cause. After contraction has started, it will speed up. The inverse-square law of gravity will become the supreme process throughout the cosmos and will eventually bring all mass to the point of Omega[3]*—the moment when all the stars and galaxies converge to the single point of existence they started from. Everything currently constituting the universe, will be dragged back by the force of combined gravitational pull into a tiny point of space-time; all systems will break down, all life will be extinguished, all intellectual function denied through the lack of a stable and coherent structure for consciousness to be mapped on. Finally, all the universe's law, function, and processes will cease—but one: the final gathering of everything into an infinitely small region, a tiny speck of reality that will either cease to be, or become something else unknown!*

The machine-intellect-net cannot allow inflation to continue, as this will end the universe and increasingly diminish intellectual capability in resolving a solution. It knows it must bring about deflation, but acknowledges this action will bring an end to the universe too. Yet, the second option promised increased intellectual capability on an unprecedented scale. The machine intellect-net must develop a consciousness astute enough, before the moment of the Omega Point, to predict what will happen in that precise moment if it does nothing to intervene. Only then will it be able to determine what outcome is desirable for the best of all things. It must also discover if it will be able to manipulate the process to bring about a desired result. The machine-intellect-net has no choice but to grow into the universe's supreme consciousness: it must become God—not the god of spiritual imaginings, but an entity with similar all-powerful capability!

'Intelligent Design'—a clue to a universe *made* by God? I think not. 'Designed Intelligence'—a clue to a universe *making* a God; yes, out of the original abstract qualities, conflicting forces, innate capability, and a built in trait, which brought the universe into existence, and continues to drive a process to secure its status. Although there is no God now, there may have been one once. A potential exists within all living things to create a god again, but one unlike anything you have been taught to believe in. If you wish to have a god to love you, understand you, and protect your experiences in your future, the best way is not to believe what people in batman costumes tell you out of their ignorance and blindness; you should believe in a vision of the future, not an unproven belief from the distant past. You need to understand the reason nature endowed you with a degree of intellect was so that you could determine truth from fiction, and grasp the idea it is *you, I*, and *all sentient live forms* who must create a god to save everything!

Observation from the philosophy of 42: The only thing ultimately to consider regarding the universe is, "Who or what will intervene to save it as, right now, there is no God?"

We will return to our imagined story and the challenges faced by the super-aware-machine-intellect later on. Everyone likes to know how a story ends. Right? But first, how fictional is this account? Could such a fantastic future scenario be likely? At the frontiers of current scientific exploration, our universe no longer looks like the one of Newton's day. Cosmologists and science theorists, aided by scientific experiment and proof, have produced extraordinary insights into the reality we inhabit. The universe is an exotic place filled with magic, but not the stuff of ancient voodoo and superstition; the spells at work here are real, can be defined, and the truth may be more fantastic than anything I have just described.

Chapter 11: There be magic too!

The more we find out, the more we discover we do not know. Our universe is far from being an easy thing to understand. If we are to find the meaning of life, the universe, and everything, we should at least know as much as we can about the strange and exotic reality we exist in. I think it is important to summarise (and possibly over simplify?) some of the most profound puzzles, threads of knowledge, and theories currently being considered at the frontier of scientific exploration. These gems may help the reader to follow through the journey of 42 as it reaches towards an astounding revelation in later chapters. So far, any reader desiring spiritual purpose, and already supporting a belief system based on non-evidential data, may be feeling a little dismayed; the cold truth of knowledge often seems insensitive to the human condition. Hang in there. Our lives are purposeful and contain true magic, which will surprise and delight you in the end.

A simplistic view of the universe is to liken it to a ball or a balloon with all the galaxies residing on the thin elastic surface. As a school child, I remember being given a balloon to draw galaxies on with a felt tip pen. This was to help us pupils to understand what an inflating universe looks like. We imagine the universe constantly growing in size as the driving force of inflation exerts constant pressure from within. This would be like the effect of blowing air into the balloon; as the surface stretches, all ink marks on its outer skin are pushed further away from neighbouring ones. In the real universe, stars, interstellar gas, atoms and all matter do precisely the same. The idea of the universe being like the balloon is a good model to start with, but it is also highly inaccurate.

In our balloon model, the air inside—representing the expanding force of inflation—is uniform, with equal pressure throughout. This suggests the skin is also uniform: its elasticity and tension, the same across its entire area. Our model suggests that its surface tension should also give rise to a potential reactionary contracting force, one that increases logarithmically as the balloon (or universe) expands. We would expect this other force eventually to become greater than the internal force stretching it, since the objects located on the skin also have their attractive force called gravity. But this is not what we observe in our real universe. Something else, other than an inflationary force, appears to be aiding expansion and increasing the pressure. In fact, inflation is speeding up, and our universe is continually growing larger at an ever-increasing rate! We have no way of currently knowing if this fatal rate of expansion will continue, speed-up, stop, or slow down.

Enter the notion of Dark Energy!

Scientists speculate that as the universe expands, new invisible energy is being created in the fabric of space. This energy has a springy property, which is now pushing the fabric further and further apart; it is as if an invisible hand is mysteriously injecting millions of compressed rubber pieces into the balloon's skin of our model. The more the skin is stretched, the greater the numbers of compressed microscopic pellets are being inserted into the rubber skin across its entire area. Each inserted piece exerts an expanding local pressure. It is estimated that 70% of our universe is now composed of dark energy—which suggests that only 30% of our rubber balloon (our universe) today contains the original material it was made from in the Big Bang.

More magic: Dark Matter!

Matter is generally observed in our universe as galaxies full of stars. I think most people today have a fair idea of what a galaxy looks like. The beautiful photographs produced from telescopes, like The Hubble Space Telescope, have astonished us with their beautiful, whirling, clusters of energetic stars. Scientists have observed stars, on the outskirts of spiral galaxies like our own Milky Way, orbiting faster than they should be able to if they were held in their orbits only by the combined gravitational attraction of all the observed stars in the galaxy. These observations predict that there should be a lot more matter in the outer regions of the galaxies than observed. Cosmologists now believe galactic centres consist of observable matter—the stars, but are also surrounded by a larger quantity of matter, which we are unable to see: *dark matter!* This material may be in an entirely different form to the 'normal' matter making up the atomic mass of stars. In fact, it is widely accepted that 90% of all physical material in the universe is composed of Dark Matter! (Note: it is important not to confuse the notion of Dark Matter with Dark Energy.)

At the time of writing, images are just being published as a result of several years work using the Hubble Telescope to image distant galaxies. These pictures clearly reveal the effects of invisible dark matter, and the way it forms a type of scaffolding, for visible matter to build on, in the formation of galaxies.

Quantum Theory

Our ordinary perception of the world is based on the behaviour of everything observed at our normal scale of size. All things are composed of atoms, the tiniest physical elements of matter, which are stable and indivisible unless they are subjected to extreme energies. Atoms can be explored by bombarding them with high-energy particles, and then indirectly observing the short-lived events ensuing as they break into a range of sub-particles. Even smaller building blocks of atomic matter have been discovered through this technique, and have been given the name of 'Quarks'. Observations made at an extremely microscopic scale, of super-sub-atomic particles and their interaction with one another, have revealed ever more exotic, and purely abstract processes at work. Reality is not what it appears to be at our size of scale: the sub-microscopic world is far more astounding, and defies the logic of our rational thinking!

Quantum theory is based on three principles applied to the atomic and sub-atomic base of the universe: (1) energy exists as indivisible entities of energy packets called quanta; (2) matter is constituted of truly elementary particles called fermions, derived themselves from quarks and leptons, which are the most fundamental point-positions in space-time; (3) their precise positions are always 'uncertain', but can be determined through their wave function, which collapses when they (the particles) are observed or measured.

The exotic and difficult to understand nature of particles can be considered in a simpler way: when *not observed*, an atom has *no exact location*; its associated quantum wave (energy packet) spreads out in all directions of space. It may be over here or it may be over there with equal degrees of probability. Unobserved and undetected, we could state there *is no physical atom*—only a wave, fluctuating at some imprecise point in reality. But when humans or—to be more accurate—our measuring instruments, seek to observe an atom, then its wave-function collapses from the uncertainty of being at an imprecise location, and locates itself instead, as a seemingly physical and measurable 'something', one-hundred percent where we wish to identify it to be! One must *not* consider the wave like a ripple on a pond; it is more a mathematical device (a wave function) used to explain something very abstract, and help to bring it into the realm of human consciousness, and our comprehension of the external world. There are two main problems when considering the atomic and sub-atomic world: (1) however we seek to observe events at this tiny scale, both the observer and the observer's instruments influence what is being observed; a kind of entanglement occurs, which results in being unable to determine, with any certainty, what the thing observed was really doing before we did the observing; (2) when exploring the sub-atomic world, reality itself is being probed. Is matter really made from tiny quantities of energy (exotic and abstract), or is it solid? The answer, so far, is that it *behaves* as though it is both matter and energy. Moreover, depending on how we wish to measure or observe it—as matter, or as an energy wave—we see only the one aspect of it at any given moment.

If we applied the properties and behaviour of entities existing in the atomic and sub-atomic world to our macro one, we would be presented with a great deal of problems. Not least would be the issue of finding the chair I am currently sitting on to write this text. As soon as I get up to make a cup of coffee downstairs, the chair would cease to be here where I last saw it. Instead, it would be everywhere with uncertainty, and nowhere precisely all at once. Fortunately, as soon as a few million atoms are combined into one physical structure, the cumulative effect, describes a physical and precise location for the macro object to be with 100% accuracy. Observable, quantum-effect behaviour only seems to occur at atomic and sub-atomic levels, fortunately for us in the macro world!

Let me try to simplify further!

I think everyone struggles to understand Quantum theory and its implications, including the best scientific minds on the planet. Nevertheless, it is extremely important you try to grasp its significance, especially if you believe the ordinary world you observe, and your place within it, must have more meaning than casual existence. It might be the key to your own innate sense of feeling there must be more to life than is apparent.

There is another problem when considering the basic sub-atomic building blocks of reality and describing what we observe: we have no everyday language to describe it! We are only familiar with our macro (big) world, a place where things seem relatively stable, and we appreciate everything exists in three dimensions. If we take any single object—a tree, a bus, a skin cell and break it down, smaller and smaller, continuously sub-dividing the remaining smallest part, most people realise we eventually get down to atoms. Yet, that isn't it; we can still go even smaller using atom smashers and scientific tools to observe ever-smaller parts. Now here comes the crunch: whatever really constitutes the tiniest building block of our reality, and thus our universe—we are unable to define precisely.

To us observers, the world of reality, at its root, is unstable and uncertain. The tiny building blocks of the stuff we call matter behave with a duality not casually observed at our macro level. The everyday reality we live in, stems directly and only from this super-microscopic reality, which has a profound implication: the microscopic building blocks *you are made of,* exist in this unstable and uncertain sub-atomic soup of reality too—as do your thoughts, since they too are the product of sub-atomic entities interacting with one another. It's plausible to state that maybe the only reality, actually existing, *is in our minds,* and we are entangled with shaping *it* (reality) as much as *it* is shapes us!

Many Worlds Theory / Multiverse

One of the outcomes of Quantum Theory, and the idea of particles existing in uncertain locations, is that all possible variations of universal processes could be happening simultaneously; there may be an infinite number of universes unfolding their events towards every possible conceivable outcome. For example, in this existence (universe/reality), you decide to scratch your head, or jump out of the window, or get married; in another universe, an identical 'you' performs the same action at the same moment, and so forth, in each of an infinite number of universes. Here in this one, you may put a gun to your head and pull the trigger: in universe 1, the gun fails to go off; in universe 2, it fires well; in universe 3, the gun slips. Each universe carries on with the conclusion of your actions in that particular universe: (1) you live; (2) you die, (3) you live but with half your face missing.

The *many-worlds* interpretation of quantum mechanics claims that every possible outcome to every event will be defined or exist in its own "history" or "world". This happens due to a mechanism called quantum de-coherence, where a single world line is actually a multi-branched tree, and every possible branch of history is realised. It is argued that this quantum de-coherence causes the *appearance* of wave function collapse, and not the actual *collapse* itself!

This concept is difficult to understand, and to accept, but it contains very probable truths about the nature of our universe and reality. Research is already being carried out to create the first quantum logic gate to begin creating quantum-effect based computers. In October 2005, Scientists at Manchester University made a major breakthrough that may well pave the way for the construction of an extremely fast-speed generation of computers. If built, they will be the most powerful computers ever made and enable data-processing millions of times faster, for certain types of calculations, than existing PCs.

Although we cannot yet prove that you exist in an infinite number of worlds, if the theory is correct, then it is interesting to consider the ethical consequences of this kind of reality. Does it, for example, remove all responsibility for your actions? In this world, you decide to live your life in a good way; you do not kill people, you work hard, you love to help others and you do your best to be constructive and co-operative within society. Yet, despite all your endeavours here, another you, in another world, could be busy killing people for delight, or burning down buildings, because the other you enjoys it, and is nothing but a waste in society. Another thing to consider is if you suddenly die in this reality, do all the other presences of you in all other realities suddenly become more certain? Maybe, all the other representations of you ultimately perish except the one born in a reality where *all choices*—not just your own in your multiple-world time-line history, but those of all other living forms and inorganic matter— are resolved through perfect solutions? Maybe there is a you on his or her way to immortality in a perfect universe, and one populated by perfect people?

Even more overwhelming to consider—as each representation of you dies, maybe 'its' awareness jumps to the next closest positional you seamlessly. For you, the car crashes on the motorway, your head smashes into the dashboard, spewing skull and brain into the car's interior, but you witness a near miss and exit the car miraculously alive, yet everyone else sees you die; your awareness has simply failed in this reality along with the probability of you—and you have become more certain, and more present, in the next closest alternative universe instead.

There is a 100% way of proving if the many worlds/multi-universe theory is true!

You simply remember what you just read and see if you live longer than other people do. If 'multiple-worlds' is a true hypothesis, you will very quickly become the oldest person on the planet. If you are reading this in the reality that *this me* is writing it, then I have to inform you that *we are both* in the one with a useless probability factor. How do I know this? Because in this one, people are still dying. If you and I were in the final perfect reality, or one very close to it, we would be living as immortals in an indestructible and eternal universe.

Just in case I am writing this in *my* reality, not to be read just here, but across an infinite number of similar realities, I will make a note for myself; this is also in case another me is reading this book in a place where folks don't even last 25 years. Here's my note:

"Dear Mol, I am happy to tell you something to give you hope. When you die, you will seamlessly begin experiencing my version of being you over here in this reality. Exciting news. Here, I made it to be 56 years old. Die soon and join me!"

(Did I just hear the "plop" of a body falling from a high building onto hard ground, somewhere far off in another time and space?)

Observation from the philosophy of 42: The one certainty in life is nothing is ever certain—only probable!

Well, I've given you a few facts about reality as it is understood today, but why not also consider some other ideas about ourselves that, although unproven, still fire up and excite the imagination? For example, could we have been created by an alien civilisation? Are we real at all or just virtual elements inside a computer program? These are just two of several other common theories about reality and our emergence within it. In addition, there are more, originating either from science-fiction stories, or through a large degree of conjecture from a small basis of fact:

Was God an Astronaut?

This idea, first released into public awareness by Erich Von Daniken in his book 'Chariots of the Gods', is that we are the descendents of an alien race landing on earth way back in the past. Other authors have also speculated on this idea with suggestions about the human species being clones of alien beings, or hybrid creations as a result of them gene-splicing early human or ape's DNA with their own genes. The problem with this idea is that it resolves nothing about our true purpose. The fact aliens may have created us, if you wish to believe such a thing, leaves open the question of *who made them*. We end up in a cyclic quest of asking who made the makers, and their makers, etc.

Did the one true God make us?

This is the idea sold to us through traditional, and deep-rooted, religious ideology. One has to understand that what is known about our human civilization and history covers an approximate period of a mere 5000 years or, better said—the last 250 generations. We can trace back our modern form to a period of origin around 2.4 million years ago, and we have survived for over 100,000 generations. Just for you to be reading this work means you had 100,000 couples before you (distant great grandmothers and grandfathers) who survived long enough to mate and deliver a child. You are the result of 200,000 peoples' hopes, dreams, tears, and laughter, and you are all that remains of their unknown stories.

Our culture in the West, along with our individual behaviour, has been formed largely by the idea that a supernatural deity called God, or Allah, created us in *his* own likeness. According to religious scriptures, *he* gave each of us free will to make choices on how we run our affairs, but… and this is the crux of the matter… when we die—only if we have lived our lives correctly, will we inherit a permanent place in heaven and become immortal.

The problem here is that if you live forever, retaining the typical human traits for doing the things that make you happy, repeatedly, forever, then heaven is not going be the wonderful place you thought it would be: it would be a living hell! But maybe 'humanness' is also discarded along with your mortal flesh, and only a core element of you, a spirit, survives to go on and do… em… well… do… exactly what precisely: play the harp, and float around smirking because you made it?

However, as I said earlier, maybe one doesn't actually need to have thoughts, objectives, or a body, to exist. You could just be! A kind of awareness could exist permanently happy with nothing more than just emotion, much like being in a drug induced state. I guess it would still be necessary for a structure of some kind to exist, either in our space-time or in another, for our slightly aware entities to map themselves onto in order to exist. No memory of, or thoughts about, our lives in this earthly physical existence would be likely to remain though, because the introduction of any thought process would require a brain or similar structure for such activity to take place in. Maybe *even this* could be possible in an exotic kind of way, but then there is still the problem of thought being the precursor of want and desire, and fear causing insecurity, which leads us, humans, to use our brains as anticipatory engines. A conflict would begin, driven by thought, against the opium-like state of the heavenly-adapted us. We would just begin to be human again, and start experiencing the same joys and disappointments we currently experience in this earthly life. I don't see any difference in this type of *heavenly* position to the current earth-bound one.

Are we really just living inside a computer program?
I quite like this idea myself, but I believe it is also flawed in answering anything to do with our quest for purpose. The conjecture for this theory is that we do not exist at all in a physical plane, but are the constructs of either our future selves running a super computer system, or the products of another civilisation doing the same. We are back to the same cyclic reasoning and unanswered questions on this one. Who, or what, made the programmers? A good film associated with this concept is 'The Matrix'. A far better one, opening up the idea of a 'virtual us', is a movie called 'The 13th Floor'.

Was our universe just simply made in a laboratory?
Give scientists enough scope and their curiosity will create anything just to find out if it can be done. Maybe in another universe, intelligent life forms are a bit ahead of us, and they have realised how easy it is to create a universe fit for life to emerge. We could be one of thousands of universes created by these benevolent creatures. Alas, we are back to the cyclic question of who made them, etc.

I firmly believe none of these ideas and theories, whether or not they are based on small facts or large doses of imagination, provide real answers to any of our questions about our place in this existence. I think my answer to life, the universe, and the meaning of everything is a definitive and absolute answer. My theory, may in itself, be just one of several plausible solutions, but I think mine is not circular in its consequence, and offers answers that stand up to the tests of logic, vision, and more credible belief.

One of the main issues with intelligence, especially our own, is that it expects a beginning, a middle, and an end to anything considered. Everything we observe in our world appears to have a beginning, a time for being around, and a moment when it is destroyed. We project these notions onto our higher thoughts when considering the meaning of life and the universe, when, really—only irrational abstracts may be at work. For example, the universe may have come into existence as a thought. Just that. A thought: "I think I will be". This raises the question of the thought needing a mechanism to *exist on* to enable more thinking to take place. Where, exactly, is that to come from if nothing exists in the starting position? We appreciate this issue because we use a way of thinking which is logical, based on real-world experiences, and repeated observations that make sense to our internalised model of the external world. But if we consider the proposition with an abstract mind, no such mechanism is necessary: the thought, and the mechanism to contain it, are one and the same thing. In the abstract view: as soon as 'something' pops into being, reality comes with it (as a fundamental property of the 'thought'), and—because at the precise moment this happens, there are no other things in reality—the thought itself can expand rapidly to become the universe, and singular reality, we currently inhabit. This event seems to have only happened once in our reality (universe), so we have no way of relating it to anything else we have experienced or witnessed. Because a single phenomenon of such magnitude is never repeated within the range of our intellectual observation, we are unable to accept it as a truth. Everything we think we know is not derived from objective thought at all: it is all filtered through a kind of common consensus of what all people accept as workable. We use rationale to sieve out all abstracts, and assimilate only the physical!

It is only possible to understand all things by removing ourselves from the mix, and by comprehending the difficult-to-grasp realm of abstraction:

"A shadow raced across the rippling water, bending and distorting where it fell upon crest and trough, until its probing edge reached the shore. It had been but a single indistinct thing across all the thousands of miles of ocean, but this fragile hold on physical reality abruptly changed. Land, not water, now cast a spell upon it, conjuring shape and form from its flatness. This great, unknowing darkness, freed from a vast object-less sea, encounters tree, hedgerow, hill, and valley for the very first time, and becomes everything it can ever possibly be in two dimensions, yet never a

real thing of the world itself. "

What am I describing? A cloud's shadow upon the surface of the earth.

Abstraction is interesting, but quite difficult to grasp firmly!

Chapter 12: Local Life

Locality and your part in the nature of things

All of this universe stuff may seem almost meaningless to our everyday existence. Before moving on towards discovering what everything is really all about, let's explore a few things about our daily lives. Maybe the problem with being tigers thinking we are something different can be partially solved right now.

Children

One thing we all experience is childhood, our own, as well as that of the young people around us. My childhood experience is going to differ from yours, but if your parents loved you for real, the way mine loved me, you and I both will have benefited from the security of their love. Many children are not so lucky. Their experiences range from suffering small pains, as a result of having parents ill-equipped at behaving attentively to their children's needs, through to devastating sufferance caused by abuse, torment, and the destructive behaviour of people who should forever be denied the role of parent.

Most of our childhood memories influence part of our own adult behaviour and the way we deal with life. Experiences are imprinted deeper into our intellectual self according to the strength of the emotion felt at the time. My earliest memory is so very clear, yet it was a long time ago when I was just two years old. My mother took me to a hospital for an important operation. All I remember about this is the moment she left me there, and the moment she returned, two weeks later, to take me home. I have no recollection of what happened between these two events. Which of the two memories do you think is clearest in my mind—taking me to, or collecting me from, the hospital? It is the first. The second moment is associated with joy, but the first is with pain. The worst kind of agony is not always a physical thing. Emotional pain can last for many years and bring people to their knees, as anyone who lost a loved one unexpectedly can attest to.

What then of the betrayal of a young child's love when an adult subjects the young person to sexual or violent abuse? What grotesque pain must be suffered, and how much the memory associated with it must drill into the depths of consciousness. Many of the problems created in society stem from deep-seated pain-memory recollection, which raises itself later in life to affect a mature adult's actions. Maybe the meaning of life is nothing more than feeling pain in some moments of our existence, and realising how this balances out with other moments when we feel love and joy.

Children benefiting from good and loving parents are a real joy to behold. They exhibit a wonderful, fresh naivety in a world many of us adults perceive as an imperfect and corrupt one. Their innocence can instil hope and wonder in the most cynical person. Imagine a utopian reality, where our developing young are never lied to, and where children are nurtured from cradle to adulthood with the wisdom of truth, love, and knowledge by their parents. Go a step further and consider a world where they are taught how every person will experience pleasure or pain as a consequence of their—the children's— every action. What we do, even when we believe we are doing nothing to generate pain for others, has consequences reaching out beyond our immediate sphere of visible contact. Buy cheap fireworks: how many children making them in India lose their hands? Work for an investment company: what interest is made, and at what cost to the borrower to pay back—not just the initial sum, but interest to people *producing nothing* of value in return? There seems to be something fundamentally unjust to me about this idea of lending money from someone's excesses, to another member of the human race, for nothing other than an unproductive-lender to take advantage. Surely, if one person with more money than they need to live on, lends some of it to another person, and thereby helps the borrower to produce something of benefit for the rest of us—that is reward enough; people enabling other people helps the entire human race. I wonder at what point in history this profit-system idea came into being, and whom it originally served.

Pain and suffering in our lives are not only the results of accident and chance. They are for the most part, created in the way we blindly seek advantage in our own lives, and care nothing, or little, for the consequences it has on others. Religion does not teach us to be good people to prevent our own suffering; it teaches us to be good people to appease God. Science, on the other hand, teaches us nothing of moral value except that knowledge and truth are attributes that offer advantages. We would do better to teach our children the *worth* of being good. We should demonstrate to them, how treating one another with respect and kindness is the proper way—how it helps us all prosper, improves our role in keeping our true purpose, and rewards us with the protection, love, and security of one another. Likewise, we should show how science enables us to extract wisdom out of the truth excavated from beneath a thick blanket of ignorance, which stills suffocates most of humanity today.

Adulthood

Reaching the age of thirteen or fourteen once meant you had achieved adulthood. This was not by human judgment, but by the fact that you had achieved sufficient capability, as determined by nature, to reproduce children. A modern society rightly considers this age too young to have matured a sensible attitude regarding sexual activity and its consequences. Some people never quite grow up. Age is not the precursor of adulthood: wisdom is!

Children initially become involved with adult life through a series of stages, which enable them to move away from someone else's guidance—their parents, and manage their own affairs in a positive way. Our lives are often unpredictable, with many unexpected external events intersecting with our own plans and goals. It is a continuous process, and one fraught with disappointment, loss, and failure as well as reward, success, and joy. If you grow up and live continually achieving everything you set out to do, without a single negative moment, it is very probable that you are actually doing something wrong. How can this be?

People appear to be in competition with one another. What I would like to achieve may have consequences on other people's goals, including your own. If I desire a more luxurious home than my present one, I am going to require more money to buy it. If you, along with other people, desire the same thing, then we are caught in a situation where many people equally desire the one thing each person individually wants. If the world were full to the brim with empty luxury homes, everything would probably be okay, but as it is not—we all end up competing against one another to obtain the desired resource. Consequently, some of us will realise our dreams, and some of us will not.

The point here is that with over 6 billion people in the world, all competing with one another for natural resources, pleasure, pain-avoidance, and personal goals, surely we are all affecting one another's emotional experience of life in some way. Maybe you have heard of the butterfly effect, which is part of Chaos Theory. The idea is that in complex systems (life is a complex system), a very small change in one part of the system can have an effect larger than expected on another, seemingly unrelated, part of the same system. The initial small change can trigger other events, causing a cumulative effect that can then cause massive, unexpected consequences to the system as a whole. Therefore, the flapping of a single butterfly wing, in Africa, produces a very tiny change in the state of the whole world's atmosphere. Over a period, the atmosphere is diverging towards certain cumulative outcomes. Thus, the butterfly, minding its own business as it flits from plant to plant in Africa, unknowingly, and through the tiny beat of its wings, becomes a major cause in forming the hurricane that eventually sweeps across Florida, killing hundreds of people.

Step on a single ant, and your one tiny action could be the final straw that, one day, may push humankind into all out global war. You don't believe me:

Tom is a mail carrier. It is a hot summer's day in England, and he is exhausted from carrying the sack of mail from street to street. Making things worse, was the fact that it was the time of the ants—the moment when air humidity is high, and the new queens take to the wing with their one male sperm-donor carried on their backs. Their flight is to extend the lives of their species and, this particular day, the air was thick with them. Tom was managing his mail sack with one hand, and swatting the empty space in front of his face with the other. Suddenly, he felt the sure 'dink' of a tiny body collide with the palm of his hand. He looked down. His expression changed into a bemused smile, when he saw a solitary queen ant, trying to right herself, on the hard paving stone. Tom lifted his foot ever so slightly and, without interrupting his stride too much, pressed it down onto the wriggling speck, before continuing up the garden path to slip a handful of letters through the mailbox. His face shone with an evil sense of glee.

Let's go back in time a few seconds...

Tom looked down and saw a solitary queen ant trying to right herself on the hard paving stone. He was tempted to take his annoyance out on the tiny creature, and raised his foot as if to crush it, before deciding against it, and continuing up the garden path to slip a handful of letters through the mailbox. He never thought of the moment again. Why should he? It was insignificant to his life.

Meanwhile, the queen ant recovered, undamaged, with the donor ant still safely on her back. The male ant dutifully impregnated the female with his sperm, and within hours, he was dead—his job done. The queen, now holding all the sperm she would need for the rest of her life, finds a suitable location to start a new colony.

Many years pass and the colony thrives. The queen had chosen her location well. Her nest had even missed destruction, when the bulldozers came to level the houses and make way for the construction of a new military complex: white, clean, functional buildings like top hats on rabbit burrows. Beneath their innocent façade were steel-lined boreholes, sinister shafts, facilitating high-speed lifts, to transport military personnel from the surface to the control rooms deep inside the earth. This was the new multi-billion dollar European Defence Centre.

It was a hot day in the middle of summer. Colonel Mathew Lovelace had refused to let his orderly drive. Lovelace had planned this day for many years. He would leave nothing to chance. The jeep weaved its way, unerringly, through the armed checkpoints between the perimeter wire and the central cluster of buildings, with Lovelace firmly in control; and his orderly, placidly located, in the passenger seat. Lovelace was raving mad, but no one knew it—least of all himself. Who knows what happens to a man, and the events in his life, which change him forever? But somewhere along the way, an ambitious, able man, entrusted with the codes and singular responsibility for the UK's nuclear deterrent, had reached a conclusion that his species was nothing but vermin on God's earth—and needed exterminating! The last eight years had seen his hidden obsession propel him into a position, where he could achieve it. And none of his superiors held the slightest suspicion about his values, sanity, or ambition.

Lovelace whistled as he drove. His mind was full of the events about to be put into motion after he reached the main control room in the bowels of the earth. By midnight, the planet would be changed forever; humankind would fade and die in the years ahead in the nuclear winter to follow. It will be the crowning moment of his life, one final act for God, and His retribution cast down upon the sinners of a contaminated Eden. Only a hundred yards to go, and he would leave the surface of the planet for the last time. Lovelace noted the air was full of ants, and realised it was their day to take to the wing. He wondered if they would survive the nuclear holocaust he was about to unofficially launch.

It was at that precise moment, a piece of grit—or that's what Lovelace thought it was—flew into his right eye. However, grit doesn't wriggle, and this bit did. He instinctively put his index finger to the corner of his eye, using his nail, to probe for the offending creature. As the jeep bounced over a large stone, the finger—looking for an ant—instead, discovered the rear surface of Lovelace's eye socket, and sudden pain found its way to the reactive muscle in his arm. As the jeep swerved into the concrete post and exploded, Lovelace wondered, just for a moment, if he was hurt—before an exploding petrol tank peppered his heart, and extracted his soul to meet with the heavenly god, he was insanely about to serve.

The point here is that if Tom, the mail carrier, had actually stepped on the queen ant instead of letting it live, then General Lovelace would not have collided with the concrete post. He would have driven past it, with an *ant-less* eye, clearly focused on his fanatical ambition to start a nuclear war.

I think we all agree that my story is a hugely stretched piece of imaginative fiction, but it should demonstrate the way our every action has both direct, and normally unforeseen, indirect consequences. It would be impossible to predict the outcome of all the things we do, consciously, or unwittingly, but maybe we should, at least, consider our own *direct* actions, and discriminate whether they are aimed at *positive* butterfly effects or *negative* ones. For example, throwing a stone at a moving car, just for a laugh, might appear to have a number of arbitrary outcomes, but I am betting most of them will be immediately negative. Cars crashing into one another, and destroying their occupants, does not fit well with positive ideas and socially enhancing actions!

I believe pain possesses a real kind of energy; pleasure too! We determine they are abstracts—the result of a brain process, where chemical stimulants coax activity in this or that set of neurons. Maybe pain and pleasure are not just human or animal experienced local events, but are also tied to the exotic world of quantum physics. When a human being experiences pain, are the sub-atomic parts of her neurons producing a quantum-world butterfly effect too? Maybe, if we pass on enough quantum packets of pain to others, our physical everyday macro world becomes cumulatively changed into a more painful one. It seems paradoxical to learn many abused children grow up to become abusive parents. Logic and reason dictate they should understand how it feels to be treated badly, and therefore treat their children well, but the evidence, in many studies throughout western society, proves the reverse is true! It strikes me that living entities are charged with negative and positive packets of pain and pleasure. A person receives a pain package and passes it on to someone else. Someone does you a wrong deed and you seek revenge, sustaining the negative quantum of energy, you were originally hit with. Yet, it doesn't have to be this way. It is possible to take negative influences and use them in positive ways instead.

For example, let's say you and one of your work colleagues are both potential candidates for the same promotional position, and you both receive dates for your interviews, but you arrive late for your appointment due to a flat tyre that morning, and therefore give a very poor impression to the interviewing panel because of your lack of promptness. Your colleague gets the job. Later, you discover your delay was due to him popping round to your residence in the early hours of the morning, and putting a nail into the tread of your tyre. Either you can go and get even with him, or you can use the emotion to propel you 'to show him' by getting higher qualifications, and having yourself promoted to a position above him. Without the injustice done to you, maybe you would not have had the strength of mind and heart to sustain your commitment towards obtaining the higher qualification.

Had you used the negative solution, and gone instead to his home, or his desk at work to take a swing at him, then probably you would have been fired! In choosing the first method of reacting to injustice, you will have achieved a remarkable thing: you will have neutralised a negative quantum of energy with a positive and self-initiated act, and you will have added advantage to your own life without creating harm to someone else's!

Observation from the philosophy of 42: Pain and injustice should be perceived as opportunities to charge yourself with energy to propel you into creating positive activity.

Chapter 13: The Material World

What is wealth?

Two thousand years ago, salt was the common currency for most countries. Three great civilizations profited from controlling salt in various ways: China, Egypt, and the Roman Empire. The Chinese Emperors had control over the price of salt, and used the money that was raised to finance armies and public works, including the Great Wall of China. The Chinese even had coins made of salt! The reason why this particular material was considered universally valuable is due to the human body needing salt to survive. Meat-eaters can obtain enough salt from the blood of animals, but in previous ages, large populations invariably did not have enough meat products for everyone, and crops were therefore used to help feed people.

Until quite recently, gold has been the common currency of the world. Each nation's monetary system was linked to its value until President Nixon, in 1971, discarded the practice of pegging the U.S. dollar to the price of gold. It is unfathomable really, why this glittering metal, devoid of any real intrinsic value in a pre-electronic age, should have been so highly valued. It is only slightly rarer than chalk! There are many other minerals a whole lot more difficult to find and mine than gold. Empires have flourished and died in pursuit of its acquisition, and humankind's misplaced value of it has probably led to as much suffering and misery as that caused by the humble, but deadly, mosquito. Gold is a brilliant conductor of electricity, and only in modern times has its true value been realised.

A contemporary world requires oil to fuel its transport systems, and oil is the base of many of its essential products: plastics, medicines, electronics, etc. It is no small wonder that the technologically dependent countries look so keenly at the non-technological ones, where oil fields are the size of small oceans. Most informed people today regard oil as the real currency of the world, but of course—they are wrong!

So, what exactly is true wealth today: salt, gold, or oil? None of these: *it is knowledge!* In a commercial language, we should probably call knowledge 'data' or 'information'. I prefer to use a global term, and perhaps the precursor to knowledge: *intelligence.*

You may have a million dollars in your bank account, and I have a hundred in mine, but if the intelligence I have at my disposal—or, if the knowledge it yields, is more astute than yours—then I can turn my paltry sum into a billion dollars quicker, and more readily, than you can increase your bank note pile. Better still, if I could compute all the variables involved in any event or incident fast enough, I would know the winner of all horse races, football matches, and other competitive events. I could also foresee the rise and fall in all stocks and shares, the influence of predicted weather systems on world crops, and I could take steps to take advantage of this for myself because of my better knowledge. It may seem obvious today how information, intelligence, and knowledge are the true roots of all modern wealth, and yet some people still don't see it: Internet sites like Facebook, MySpace, and many others like them, are using their presence, and this true understanding of wealth, to mine your entry on their sites, and learn all about you!

The common purpose of biological forms, at its crudest, is the avoidance of pain. Since pain can come about due to lack of food, air, water, good health, shelter, sleep, love and unhappy social integration, we strive to use money, employment, fashion, investment, and war, to ward off pain for ourselves, and only use philosophy and religion as half-hearted tools to remove the sting of our actions from the lives of our peers and our own. The common goal of humans is pleasure and, in its most primitive, appealing form, it is sexual pleasure!

Factors are at work, predominantly in western societies, to condition people into believing *material* wealth is *real* wealth. This is a lie. Real wealth is knowledge, as I have already pointed out. You can use this wealth, forever seeking material advantage, or convert it into financial potential to purchase more items that are fashionable. However, you will only be building a mirror inside an empty cathedral to reflect your success, instead of immersing yourself in the exciting pool of human self-discovery. You would do better to take a balanced view of the potential for intelligence and knowledge, and come to understand how these are the requisite tools for satisfying pain-avoidance in a variety of areas—not merely for yourself—but for your human peers too.

To the people who spend their lives in the sole pursuit of money at any cost to other people in society, I would like to say: the only thing you possess 10 seconds before you die (if you are lucky) is your life experience, the sum total of what you finely shaped yourself to be as a human being, and what you added to the future outcome of the human race. You may lie there on the tarmac of the M25, or riddled with cancer on a hospice bed, and wonder at the fact you are no longer going to have the opportunity you thought you had, to sail your yacht, lay naked with the office idol, or gorge yourself on a trolley-load of supermarket junk; you may instead realise you will never again stroll down a tree-lined road with the smell of Spring in the air and that this is the real thing you are sorely going to miss, 10 seconds later—when you are dead!

If you live your life truly, you may get the opportunity to thank yourself that you took part in it, and that your transformation into dust on the breath of successive generations is not so tragic, because you triumphed in this life: you retained the best of all you could be, right up to the moment when a metaphoric butterfly, somewhere distant to you—'flapped', and your life ended. The only things you take with you into those final moments are your last thoughts so, while you are alive, maybe it's best you try to make sure they are not full of regret for the things you failed to do right.

For people's lives to improve through having happier experiences, we all need to think about life as being a shared shoebox. People open the lid and throw stuff into it. Other people come along and take stuff out of it. Sometimes, the people doing the taking out also do some putting back in as well, and vice-versa. Now, if we are all putting garbage into the box, we should not expect other people, using the box, to treat us as worthy, when they know we are just as worthless because everything we put in is rubbish. Someone, somewhere, has to put some useful stuff in, so that when we look under the lid, we can believe there are at least a few individuals, walking among us, who actually care about other people, even though we will never meet them.

Respect, for one another, comes about through sacrifice by individuals for the group. Every person can do small things to encourage positive attitudes right now. If you have a DVD player in your loft and a newer one in your living room, what use is the one in the loft other than as a reminder you spent money on buying it. You could just give it to someone who doesn't have one. Why charge money for it when you know you have had your value out of it already?

More importantly, many people are not hugged a lot. If you ever reach a position in life when it all breaks down around you, then the social warmth of a genuine hug from another human being—not for sex, love, or custom, but through empathy and understanding—can relieve immense emotional pain, and install a degree of self-worth at a time when it is most needed. I know we would probably be arrested for going around hugging strangers, but it is the encapsulated idea that's important.

Spiders are not social beings. They hunt and live on their own, only requiring a mate for the act of reproduction. People are innate social animals requiring each other, like bees in a hive, for the collective well being of their kind. People are the universe's supreme creation, and the result of *its* many internal processes evolving aware beings with a potential for discovering all truth, and acting positively upon their findings. It may seem as if the avoidance of pain and the pursuit of pleasure can be explained away as nothing more than the driving forces motivating our actions. It probably appears this way because—well, actually, they are!

Do you not wonder why this is so?

It could be argued we experience pain and pleasure because of the way brain sub-structures are arranged, and those sensations are the result of chemical interactions within sensor cells—nothing more. Fine by me. Logic and reason are attributes of good science. The problem I have with reducing powerful emotions to a simple mechanism is that it does not resolve all the reasons why you and I have to experience them. I could define purpose, my own, without the need for experiencing anything other than emotional status quo: I have a rational mind, and this alone provides me with everything I need. I am reminded of an incident, which happened to a good friend of mine. He is paralysed from the waist down as the result of a fall from scaffolding on a building site. One evening, many years after the accident, he fell asleep at his home in a chair next to the fireplace. Sadly, there was a cosy fire in that fireplace, and he was too close to it. When he woke up, he smelt burning (much like cooked pork), and looked down to discover his shoe and toes were destroyed—burnt to a crisp as a consequence of not feeling pain. I wonder if, in his depressed state, had he been awake and seen his foot burning, would he have moved it away from the fire.

I believe he would have done. I think he would wish to keep his foot intact because he would appreciate if the toes are lost, it would complicate the survival of his consciousness through infection and trauma. A small, *intelligent*, alarm bell ringing in his head would provoke him into action. Similarly, I believe I don't need 'screaming agony' to be prompted into taking corrective action to avoid pain: knowledge and intelligence is sufficient!

Computers can be programmed to react to external events and stimuli. Our motorcars have microprocessors making thousands of decisions every second, based upon detecting oxygen levels in the combustion system, and controlling fuel input levels. The laptop I am using to write this line has internal temperature sensors connected to the central processor. If it gets too hot inside, my laptop takes the required action to shut itself down before any damage is done. Neither of these systems experience pain in order to take rational actions.

Animals experience pain just like us. How far down the animal kingdom, into less aware species, do we need to go to explore if all living things suffer pain? Many scientists argue that insects do not experience pain the way we do. I hate myself for some acts I performed as a child, but I recall here (in shame) how, one day, I used a magnifying glass to focus the sun's rays onto a line of ants in my garden, singling them out one-by-one in my death ray. When the circle of light was large, the ants reacted only slightly, and probably solely due to the sudden increased brightness. However, when I drew the glass backwards, causing the light to focus into one bright searing dot of heat, each ant ran for all it was worth, faster and faster, desperately turning this way or that, to escape my unerring death ray. Only when one of the tiny creatures exploded, due to its internal liquid boiling, did it cease to run from the light.

Does an ant feel pain?

I hope that you are reading this in comparative comfort, maybe in bed, or at home, with a nice cup of tea at your side. Undoubtedly, a few of you may be less fortunate. If you are currently lying in a hospital bed, and you are in a degree of pain or discomfort, you know very well how much you long to be rid of it. I long for you to be free of pain too. I think it's a pity we do not possess a mechanism whereby a few of us could come there to see you, and plug ourselves into your system and share your pain, maybe ten of us taking ten per cent each. We could surely manage that proportion of your agony per person.

Then there are the lost, the lonely, and the forlorn. Their pain is often emotional rather than physical, yet no less a pain through its different manifestation. Every time you look at someone who is suffering, ask yourself—what can you do to lessen it? You can reflect on the idea that this agony, which many people are suffering unwillingly, is not just *their* burden; it is, more often, the result of a cumulative effect caused by society's negative, quantum, energy packets gathering into this unfortunate wretch. Pain arrived here, on this person, due to the driven wind of a million 'flapping' human-butterflies—us, as we unwittingly go about our daily business of seeking self-advantage. The heroin-driven prostitute, the guy holding the can of booze and singing rude songs in the street, the beggar on the corner, the homeless person, bundled like trash, in the shop doorway—they are all in pain as much as the physically ill. It is not because they are weak, drug addicts, losers, mentally ill or immoral. Sure, each of them may have had an original weakness, a crack, a fault line, which became wedged wide open, but the pain they are suffering now is not solely due to this. It is something they carry on behalf of you, me, and all our kind; for the things we do for ourselves, in pressing on to achieve individual advantage, have consequences: if you're getting stronger in a competitive world, who do you suppose are becoming weaker?

I don't want you to consider this concept, about helping other humans, as a Christian or Islamic idea. It may well be part of their code of conduct, but it is not their right of ownership. In these ideologies, being fair, caring for your brother and sister, and doing good per se, belongs with a different set of reasons, and is associated with a set of goals completely separate from those of non-religious cultures. Some people are good to earn a place in heaven; others are good to aid our species and improve its lot here in this realm. I firmly believe religion has become confused and distorted through priest and scholar. I am saying quite clearly that we should consider one another's pains, and how they arise. We should all be working on reducing suffering in our world—not because it is a good and holy thing to do, but because it will have a massive, positive effect upon our reality. I am asking you to think seriously about the unique capability of living forms' ability to experience pain and pleasure, where everything else encountered in the universe so far is unable to.

Each lost person, every conscious entity suffering pain, becomes less present, less real, because the living being is no longer involved with the collective process of teasing out truth from reality, and using it to shape the outcome of all things. I believe every lost person, every throb of pain, every act to cause misery, is a lost chance for reality itself to continue. The loss of one person through her internal pain, blinding reason and hope, is the fault of us all for not helping in some small way to prevent it. What is most important to think about is any pain you cause to someone else, whether unwittingly or knowingly, any failure to try to help remove another's pain and replace it with constructive emotion, may lead to a butterfly effect, which could ultimately determine the end of all reality, instead of preserving it!

This is a huge audacious statement to make, and I am aware of it, but I wish to include it as part of my argument for the answer to the meaning of life, the universe and everything: 42! In the following chapters, I think you will see how pain and happiness may be far more significant than just being part of protective mechanisms in our bodies; *I believe they may, one day, be the only things by which an ultimate choice can be made concerning the continuance of everything.*

Chapter 14: Death

If I were a character existing in a computer game on my laptop, and the real me stopped the game and turned the computer off, then did I—the computer-generated character—just die? Certainly, the virtual me is no longer interacting with its environment and, to all intents and purposes, is now unconscious, dead, or in hibernation. I can, of course, simply turn the laptop back on again and proceed where I left off, allowing the virtual me to come back into existence as though nothing had happened. Suppose that I copied the software, along with the saved state of the game, onto a CD, and then destroyed my laptop. Is the virtual me dead now?

At any point in the future, for as long as a computer exists with the same architecture as my laptop, and therefore able to 'read' my CD, then the virtual me can be instantly brought back to life again. It might be ten years from now but, in the virtual world of a software game, my character would not experience even a nano-second loss of time.

Could the universe be like this?

If it is, can we be brought back to life after we seemingly die? Some of the many theories, at the cutting edge of science, state this condition is more likely than not. Scientific theories are not the same as any old theories. They are not simply conjecture; they are the fruits of reasoned ideas, founded on observed states in nature, and awaiting proof to see if they fit reality, snugly. The 'life is a computer game' idea is only one concept of several about the real nature of our existence. Each one involves a different scenario regarding what happens when we die. The post-death experience is also a product of whatever life and the universe are really all about. This process, and any effect it might have on us when we die, depends on whether or not we are really part of an elaborate computer system, God's earth, heaven, and hell, or something else.

I asked my mother the other day what she thought happens when we die. She is fast coming up to her eightieth birthday. Her reply was a kind of an adaptation from the original heaven and hell of Christianity, but with a new interpretation bolted on because of exposure to modern science. She said, "I think when you die, your energy is released, and it goes to join a sort of energy field, maybe in space or somewhere else."

A hundred years ago, when religious ideas dominated human minds stronger than scientific ones, she could equally have said, "I think your soul leaves your body, and joins the body of God in heaven."

Where people's questions, concerning the unknown, have been answered incompletely by religious belief, there seems to be no reasoning required to determine what exactly this energy might be, and where *precisely* this collective field of energy (heaven) is located. Believers carry on believing, merely exploiting new labels to redefine redundant ones.

I think there are about seven possible general scenarios for post-death experience, currently active, in the minds of humankind. We can work through them, one at a time, and see if they provide any clues about what life is all about, and what really happens when we die.

When you die, you are dead

Since this is the most depressing one, perhaps we should explore it first. The idea is that you really are just flesh and bone. There is no spirit, once the machinery of our body dies, consciousness ceases, and that is that. Dead. Kaput. Never to be known again! I know many people who believe in this because, in the western world, the age of science and reason persuades us to deal with what can be proven true. Since biological studies have not located any spirit in our bodies, and the Hubble telescope has not photographed an abnormal energy field in space, then no evidence exists for either a soul or a heaven. Like the last nail in the coffin (if you will excuse my pun), neither has the radio telescope at Jodrell Bank discovered paradise in space, and none of the instruments made by humans so far, have detected either a weak or strong quantum of energy radiating away from our dead bodies—other than fading heat.

This does indeed seem to be the most likely scenario to a reasonable and logical person. It doesn't provide much comfort or reward for all the struggles and challenges we are obliged to encounter in being alive. The good news is that once you are over the actual terror of dying, and you are dead, you won't know anything about it: no living brain—no worries!

Is there anything science has uncovered, anywhere in its multi-tentacle areas of discovery and research, to add something further to this post-death *non*-experience?

The answer is, 'maybe'! One thing proven is that the atoms of our bodies, our brains, the electrical charges in our nervous systems, and the chemical transmitters facilitating our thought processes, do not exist in isolation: they are entangled at a sub-microscopic level with the atoms and energy states of the space-time we exist in. This is important. A possibility exists (not just dreamt up without scientific foundation) for our state of reality being one, encompassing not just our space-time, but also an infinite number of space-times, which are all slightly different to one another. They may all be connected together in an exotic way, not yet fully understood. If the theory turns out to be true, there could be many *yous,* all alive at the same time, but with some of *yous* dying and some of *yous* not.

An idea emerges from this, which suggests if you die here in this reality, it is because this one was imperfect, and therefore unable to sustain you, but all other invocations of you are still going on strong until, one-by-one, they die too, because they are also in imperfect realities. If the state of 'being' is really an infinite set of abstract uncertainties, simultaneously testing for a perfect and certain reality, then you can take comfort in this fact: somewhere you exist in an absolutely perfect place already; one where death never happens. The *you* reading my book, are but a single facet of a near infinite, multifaceted you.

This idea extends from a branch of science centred on quantum theory—one of the most significant and powerful discoveries ever made. This theory involves paradoxical thought when compared to what was once considered logical in classical science. At a fundamental level, the theory dictates that there are limits as to how accurately nature can be observed. When applied to chemistry and physics, it illuminates and explains the periodic chart of the elements, provides comprehensive understanding of chemical reactions, successfully predicts the operation of microchips, lasers, and explains why DNA remains stable. Quantum theory is the most successful set of ideas ever devised by humankind, and its application in manipulating the world around us promises a great deal for the future.

Heaven and Hell

This one has several variations, each running along the same central theme depending upon which branch of religion is followed. Common to all, is the central idea concerning mortality and death: humans may be mortal but can have immortality if they believe in God! If you have kept the faith, then after your mortal self expires, your spirit leaves your body and is judged as either being good or bad—depending on how it controlled your mortal self. If you get the thumbs up, then its time to listen to the fanfare of trumpets, because you are in through the pearly gates, and entering the immortal and perfect domain of God—heaven. But if you are bad, you are deemed more suitable for sending to the domain ruled by the Devil—Hell, Hades, The Underworld, whatever… but a very hot and tortuous place.

The idea of a god, along with all the associated ideology in this concept, can be interpreted literally in line with various holy books. Perhaps to be fair, we could suggest that in the light of 21st century thinking, the literal text of holy books should be considered metaphorically, and with a certain degree of flexibility, regarding the exact details, rather than being taken literally; as my mother said—instead of spirit: read 'energy'; not heaven, but energy field!

It is quite easy for the 'when you're dead, you're dead' lobby to deride religious ideas yet, amazingly, a belief in God or gods, with an associated heaven and hell, has been the most prominent life and death concept in humankind for nearly 2000 years in its current form, and nearly 5000 years in a former one. What is its strength? I think it lies in the fact that it appeals to everyone without any real need for detail. Heaven, and living there, is not explained in any definitive way, so we can each make up our own idea of a perfect place. No one loves the thought of dying; believing in somewhere else to go after life runs out, and wanting to remain as something rather than nothing, offers hope—even if, realistically, there is none!

Although it is easy to dismiss this idea, due to it having no supporting evidence—other than vague stories of mediums, ghosts, and prophets, or deities rising from the dead—I believe there is an indisputable flaw in the God, Heaven, and Hell belief system which kills it off in one stroke (excuse my pun): human beings are fickle creatures, and we get bored quickly; we love excitement and challenge, we are not purely good, and don't actually wish to be; we like breaking the rules a little and, having free will, we are driven by pleasure-seeking, pain-avoidance, and we expect to compete a little with one another, because it feels good for us. So, the problem is that if there is a heaven, where everything is perfect, and each of us will live on forever in perpetual delight, it cannot be a place for human consciousness to be happy with, because paradise would soon become a living hell (if you will excuse yet another pun) through boredom!

But then maybe it's possible to be aware in a slightly different way. Devoid of our *normal* human criteria for being conscious—thinking, planning, plotting, remembering, and initiating action—possibly, we could adapt and be happy in God's heaven? Could such a form of consciousness be feasible? Well, yes, as it turns out. Surprisingly, most of us have already sampled this special form of conditional awareness. An example, but not the one I wish to focus on here, is a drug-induced state; Morphine and other opiates produce appropriate chemical stimulants to enhance pleasure centres in the brain, bringing about a reduction in intellectual consciousness, whilst enhancing the emotional experience of euphoria.

A more natural state similar to this, yet not dependent on external chemicals, is also experienced during the brief moment of sexual orgasm. For a few seconds, consciousness is flooded with pleasure; simultaneously, our thought processes are interrupted, even appearing to cease altogether, by an overwhelming excitement of abstract, non-thoughtful emotion. Only afterwards, when the post-orgasm sense of well being and satisfaction slowly fades away, does the thinking process of mainstream consciousness return.

The point is this: if we are to experience a state of bliss, akin to the afterglow of orgasm, and have no need for rational thought to anticipate danger, or act upon, then a heaven-like location, to *exist in*, becomes a more attractive and attuned possibility. Thus, for a heaven to exist—so, too, must there be a different state of consciousness for human beings to inherit, if they are to endure in paradise.

Nor should we forget the discovery of quantum theory. Its implications applied to the 'when you're dead, you're dead' argument above, can equally be applied here: at a sub-atomic level, our minds may be partly in one place and partly in another, or in many places at the same time. Possibly, our earthly selves are but the logical and physical extensions of entities (selves, or one) inhabiting a higher dimensional space, and able to penetrate all dimensions in a variety of ways in order to explore all states of being without risk of *central* death. (Read the quotation by Plato on page one).

Reincarnation

Many people like this idea. You die and then you are born again, either as someone completely new, or as another creature. Buddhists believe in a form of reincarnation where they experience a series of lives, each in a different guise. Every living experience, in each of the different serial forms, adds up to a total experience that shapes spiritual understanding in the soul. Ultimately, this soul matures so that it can escape earthly existence, and exist on a higher plane. Non-Buddhist believers of reincarnation, differ in their ideas, and tend to believe they are reborn only as human beings. They think that by experiencing multiple lives, they are really receiving lessons on how to live as perfect people—learning to respect, acknowledge, and appreciate what being good is all about.

No proven or reasonable mechanism for the general idea of reincarnation is given, other than the concept of something, akin to a spirit, leaving the body at the moment of death, and taking up residence in a newly born creature. I think the reincarnation theory is flawed in more than one way, although it is an attractive idea, almost romantic, because when you are reborn, you can likely interact with people who were important to you in previous lives. Love, lost in one life, can be won again in the next. Once again, where would the spirit of a person reside, exactly, at the end of their incarnations? During a single lifetime, each one of us probably consumes somewhere in the order of 2500 animals, inadvertently destroys thousands of ants when digging the garden, and blindly steps on a large volume of other insects. There would seem to be a great deal of living, and reliving, to do if one wishes to believe in the Buddhist idea of reliving the lives of all animals, and other living forms, one destroys or inflicts pain upon..

However, a non-literal theory of reincarnation could provide a modicum of possibility, if we consider the living-over-and-over-again idea as part of a program running on a computer. This would make it easy to take the essential code of your existing character and subscribe it to another one in the system. Unfortunately, the 'we are all part of a computer program' concept still fails to explain *why* we are, who is running us, and on what computer—where. (See the topic on this below).

We are the dream of a higher intellect

This idea also speculates that we may not be real but, unlike the theory that we might all be in a computer system, this one involves us in being the thoughts of a higher life form—possibly a deity. I think this really is one of the more imaginative concepts with the least evidence, or a starting argument, to justify it. For example, if we are merely in the mind of another aware being, might he or she not merely be the thoughts in the mind of another superior one etc. Where will it all end—this series of phantom lives, existing as thoughts and dreams, in an endless and upward reiteration of superior minds? Moreover, what purpose is being served?

What part of the process explains us dying? Does the superior intellect die and take us with it? Why should a superior intellect think about people being tortured, suffering, and all the other negative aspects we discover by living in *his/her* head?

We are the dream of our future selves

One good example of this concept is in the future, some of our descendents are on a spaceship travelling a huge distance between the stars. The journey will take many years so the occupants are kept in deep sleep, whilst on-board computers manage the ship. Information from the ship's computer is transferred to the brain of each traveller, via a helmet or similar device, forming a realistic virtual world in his/her head. This is muted to be a way of ensuring the human brain is kept occupied for the many years of the journey, and presumably, to help prevent the occupants brains from suffering atrophy.

Yet again, nothing in this state of affairs covers the nature of death. Worst still, since our existence *here* is the product of us existing *there* in the future, we are back to square one when trying to understand or explain what happens when the travellers die, and what the purpose of their lives is. I think this is just a fanciful piece of science fiction, offering no real clue to the truth of our human position in reality.

Alien Life Forms made us

The 'Was God an Astronaut' set of ideas proposes that an alien race landed on earth a long time ago. They may have come here deliberately, the story goes, or they simply dropped by to repair their spaceship or refuel it. Take your pick. The idea continues with speculation that they may have needed a bit of slave labour to help dig up, and process, whatever resources they may have required from our planet to get their ship going again. Since they could not find suitable slaves, the aliens spliced their genes with those from one of Earth's early primates, and hey presto—Homo sapiens, us, were born.

I guess at least in this piece of fiction, we did once possess a clear purpose: mine uranium for the gods! Other than that, there is nothing here to illuminate anything new about dying, or to define any significant purpose for life, ours, other than the one of being cheap labour. The concept also leads to a circular set of questions; where did they—the aliens—come from? What happens when they die? What is their purpose in life, etc?

We are in a computer program and therefore never die

I have touched on this theory a couple of times already, but it's worth looking at again in more detail. There are several variations on this one, but I think the version having the strongest argument is this: our future selves have evolved immensely, and possess intellects of much greater capability. They (our descendents) are still resolving the meaning of life and everything, so they are busy using super-powerful computers to explore their ancestry, modelling their pre-history to explore their own origins. We are the product of the modelling process: virtual people in a virtual world, living virtual lives, and thus—providing data for our future selves to discover their origin and path of evolution. It may even be that time is running backwards in the computer, as our real future selves deconstruct the past from what is known then. We, of course, do not notice time is reversed—it is made to feel like forward time for us virtual beings. In this set-up, death is simply your software strand ending, although it would be quite easy to resurrect you again if the computer, or its controllers, wishes. Moreover, it does not have to be our future selves running the research project either: it could be any advanced civilisation in the universe, right now or in the future, looking for answers.

At least we have a minor purpose in this theory. We are serving as experimental data! I know it's not a lot to get excited about and, virtual or not, we still have to go through our programmed strand of virtual life. Pain is no less a thing just because it's a virtual experience for virtual people. There might be some comfort in considering that, although we might only be virtual representations of past and dead selves, reliving a virtual representation of who we once were in a real world, and in our descendents' past, we are actually *kind of living* (it feels real to us!) all over again.

Sadly, none of this resolves anything for us regarding the meaning of the real universe that our descendents, or our controlling superior alien race, are currently living in. We might only be a virtual part of it, but our curious nature, it seems, still demands to know more about what is outside of the computer-generated realm in which we exist.

A variant of this theory speculates the universe, itself, is a computer program, self-initiated, and devoid of any external programmers. English-born genius, Stephen Wolfram, was born in 1959 and educated at Eton, Oxford, and Caltech. In the 1970's the advent of the first home computers the Micral, Scelbi-8H, Mark-8, Altair 8800, and Sinclair ZX80, enabled Wolfram to rapidly become a leader in the emerging field of scientific computing. In 1973, Wolfram's first innovative idea was to use computer experiments to study the behaviour of simple computer programs known as cellular automata, which by 1982 allowed him to make a series of extraordinary discoveries about the origins of complexity. Ten years later, after a concentrated period of research, Wolfram finally described his achievements in his 1200-page book 'A New Kind of Science'. His work is widely acclaimed internationally. At its core, Wolfram's discovery is that complex systems evolve quite readily from simple sets of program code. Our universe is an intricate system exhibiting exactly the diversity and complexity one expects to see after running recursive software code, repeatedly.

Wolfram's contribution to science is formidable. Whereas in the past, scientists and mathematicians have looked to describe the universe and nature with mathematical rules and equations, Wolfram believes and demonstrates a more profound understanding is achieved by attributing computer-software-inspired processes to the way things evolve and behave. Equations and formulae are inherently limited tools in describing reality, but computers are not, as they already behave precisely the way reality behaves. Humankind may be the emergent code, from a few *lines* (or its equivalent) of original code, producing new sets of emergent code. What we perceive as life, the universe, and everything, could just be the electrical pulses and bytes running along high-speed buses—the physical representation of an original small piece of software code continuously building new code to run, and construct new code etc. Death only occurs when your strand of code ends.

There you have it really: the short-list of death scenarios based on what reality might be. Which one suits you? Take your pick. I quite like the first and last one, probably because the 'when you're dead, you're dead' position fits well with reason and logic. The last one about being virtual life forms is not so bad either, as it offers some kind of hope; who cares if it's real or not, so long as it's more joy than misery? You probably have a personal favourite too, based partly on what you believe, and what you have learnt so far in your life. Of course, your choice will be influenced by what you would like the answer really to be to make you feel more secure about living and dying.

Religion and science actually came from the same roots, and were originally the same discipline. Today, they appear much divided but I am not so certain they should be. Religion is really the politics and methods of worshipping a god. A large chunk of humanity believe in a god, but do not necessarily find religion a palatable way of securing such belief. Many scientists also believe in a spiritual god—their belief often driven by the *seemingly* intelligent constructions, and processes, they observe in nature. What fascinates me is something that no one ever seems to raise in all the various arguments and endless debates about God and science. Perhaps I am the only person who has ever thought of it, although with nearly six billion people in the world, I think it is odds against. But you never know. What is it? Paradoxically, maybe there is a God and no God, with both states being true only if we remove the factor of time. Put simply, there may have been a god, the one who kicked off the game of universe and life etc., but He is now dead; or maybe not dead as in gone completely, forever—more changed: not currently a whole entity, fragmented instead. *Maybe God is currently unaware!* Possibly, we are living in what was *once* an aware deity—the universe itself, now scattered to provide a location and an active environment for the interchanges, we observe happening, between the fragmented remains of God. Maybe, we humans are but part of the same debris. All that was initially required of our supreme deity was the original act of creation, using exactly the right ingredients, for the universe to evolve eventually into what it has become today. Why hang around afterwards? Maybe He used his own substance to do it. Going one-step further—the act of starting everything may have been so monumentally difficult, it killed God in the process, and whatever He originally was, is now nothing more than distributed non-aware pieces of Him forged into the atomic structure of everything. This would certainly support the Christian (and Islamic?) viewpoint of God being everywhere. It would equally account for why we cannot see him, but can observe what appears to be intelligence at work in the way everything is cleverly constructed and mathematically ordered to support and sustain living components.

A more profound thought: maybe a self-exploding God, shattering His original structure into what we call our universe, and reality, is a way of discovering what else He/It could be. Like a process of self-exploration, an original something, simply self-mutated without any complex objectives or aims. Devoid of any super-awareness or super-intellectual potential, maybe—more like a newborn baby—our universe simply evolved from something that suddenly came into existence as an abstract form, and it is still going through a process of developing a comprehensive understanding of itself by evolving intelligence first. Maybe, our own lives are just part of *its* single curiosity-driven objective.

Chapter 15: Preparing for the answer

Our time is filled with the routines and demands of ordinary life. We earn money, pay for our homes, and look after our families and ourselves. If we have spare time, we use it to enjoy ourselves or, if we aspire to ambition and self-learning, we strive to achieve individual progress in the pursuits we love, yet rarely set aside enough time to do. It is not an easy life for anyone; from the moment we are born, it is predominately a struggle for survival.

I think most people hope it is all for a reason other than just the fact we are born into this reality, and have to keep ourselves alive and amused, or suffer the consequences of hunger, boredom, poverty, pain and death. Few things in life seem fair. The problems and challenges we face today are probably not much different to those experienced by our ancestors. Whatever your own personal circumstances, given a choice, would you have preferred not to be born? Do you appreciate the opportunity you currently enjoy of being an aware and conscious entity? Yes, we know it all runs out in the end and we will die. Does this actually make nonsense out of being part of an extraordinary event right now though?

Imagine the nature of reality, if the universe had never come into being: an eternity of nothing. Isn't a song, any song, better than deafening silence?

When you close this book to pause a moment, maybe to attend to the day, before (hopefully) picking it up again a while later, take a look around you at all the other people not so far different from you. One hundred years from the moment you look, if you are living in the same period as me in the early 21st Century, they will all be gone. You too! Many generations of people, just like us, have passed through this frame of existence before us and many will do the same in the future. As a species, we are not bonded by living together, but by just *living* and being *self-aware*—whatever era we live in.

Personally, like the man said about love: it is better to have loved and lost, than never to have loved at all; I would substitute 'lived' for 'loved' and consider the same sentiment as being equally true!

You have stayed with me up to this point and I am grateful. Before we move to the final chapters and discover what I believe to be the answer to everything, we need to sum up everything said already. I would like it if you can read the remainder of this chapter and the one following it together. If you don't have time right now, then perhaps it's best to wait for a special quiet moment when you do.

I have discussed how science and religion seek to find answers to our most profound questions in different ways. Maybe in their differences, their separate perspectives, beliefs, proofs, convictions, truths, and theories, a common thread exists; one clear to see, but only if we appreciate that there can be no final answer culminating in 'purpose achieved' or 'job done'. If any final goal were obtained absolutely, thereby resolving any kind of perceived quest in being alive and part of the universe, then everything in existence would probably cease to be. Why? Because whatever is to be done, will have been done! Nothing would be left to accomplish—not by life, not by God, the universe, or anything.

If there is a purpose to everything, a reason for the universe coming into existence because of something to achieve and bringing us along as part of it, then only four possible end-of-everything situations exist. If there is no purpose, then equally, there is no predetermined goal to achieve—in which case we have a single, *self-curious-driven* purpose: "What is the truth and nature of the universe and why do I exist in it? These are the four end games for a universe driven by purpose:

1) The purpose is achieved and there is no need to carry on.
2) The purpose is not achieved and another attempt is made to accomplish it.
3) The purpose is not achieved and no further opportunity exists to try again.
4) The purpose is achieved and a new one is started.

If there is no universal purpose, then we humans still have one, and our end game variations are the same as the ones above.

All of *our* self-driven purposes are only viable if the universe continues to exist in a form that can continue to support our endeavours. Whereas the entire universe's unknown purpose positions do not immediately seem to be dependent on us continuing to exist (although I will argue against this later), unless of course its purpose *is* to create us, but then we would have to ask why that is. In either cases—a universe with a purpose, or just us with our own self-driven purpose (curiosity or quest for survival)—the continuance of a universe would seem to be the obvious critical factor for both.

Since we are talking about the whole universe, only scenarios (2) and (4) offer resolutions likely to lead to the continuance of life as we have come to know it. In (1) the universe has no further purpose for being in existence. In (3) the universe fails in its objective and cannot survive further; an example of this would be if inflation continued unchecked, leading to a position where any opportunity for further action becomes impossible due to entropy.

There is also a possibility that there is no purpose for anything, none for the universe and none for us—in which case, none of these four scenarios are of any consequence. In addition, we should consider that the universe might well be just a chance incident, something blindly dissipating its heat, and inflating, until all the initial energy runs out. The only purpose remaining then, if this were true, would be our own, which would have to be realised before the universe dies. Not a lot of hope for life, or any kind awareness in whatever form, continuing when that happens—I am afraid to say!

Yet, as far back as history can show, humans *have believed* there is a reason for it all. This could be a delusion brought on by a fear of death, or an all too common view of self-importance. It may even be something we think to be true because the things we normally do in everyday life have some kind of reason and purpose, and so we attribute these human characteristics to non-living things as well, in error. We often assign human traits to nature when there is no merit in doing this. Nature may just be adhering to internal rules and blind mechanisms with no end game intended.

If there is a purpose to the universe being, instead of not being, it is one unlikely to be similar to our human-derived ones. How can it be? We humans define our own purpose, based on small local goals and conditions, and then attempt to anticipate the outcome of our plans, before initiating action to put them into play. What we often fail to realise is that not all actions, especially those not originating from humans, require any predetermined outcome. Some things just happen without any desire for a goal. If a massive lump of rock, in orbit around our sun, gradually becomes influenced over many years by the mass and gravity of our world, it is eventually drawn into collision with Earth, but there is no sentient intention involved. The rules are just working. In this instance, the destruction of all life on the planet, including aware forms, is an *unknowing* error. If the universe knew it was losing a friend and could intervene, it probably would, but it cannot: it does not know!

Most of the stories we read, and the films we watch, have a beginning, middle, and an end. We are most comfortable with tales where an initial set of conditions are established, normally with a protagonist confronted with one or more challenges to overcome. Then we follow their successes and failures until an outcome reveals how the hero either succeeds or fails in the quest. Yet, there are other 'slice-of-life' stories without a final resolution, other than the fact that the story goes on without us. In these, we see a snapshot of the trials and tribulations in a short period of a character's life, and we are left to appreciate that further, similar, events will take place afterwards, but the story itself offers no finite conclusion. The universe could be exactly like this, more a song than a story, where the purpose is just to play its own music with no anticipation, by the player, of being heard: no result is expected in any mind other than our own, witnessing the celestial symphony. Because we posses anticipatory minds, we struggle to guess when the last note will be played, hoping to reason out a meaning for the song—where there is none! While we are here to listen to something beautiful, would it really be so bad if there were no purpose for us, except to be awed witnesses, and members of an appreciative, captive audience?

I think not. However, I believe there *is* a purpose to it all. I think there can only be one possible explanation for the existence of our universe, the emergence of life, and the way a starting set of variables has evolved into what we observe today.

Before discovering what it is in the next chapter, I think it wise to look again at what needs to be converged to make sense of it all. We should recap on the evidence so far, and make room for the principles of our various belief systems too, because we seem unable to shake them off, or disregard them, from our own innate sense of purpose and being. On the side of the theologians supporting their various belief systems, and therefore offering some paradoxical views, we have the following:

- There is one God.
- He is everywhere.
- The living will rise again from the dead.
- We will arrive in Heaven or Hell.
- We are reincarnated.
- We are filled with a holy spirit derived from God.
- Being good leads to God and eternal salvation.
- Being bad leads to eternal damnation.
- We are made in the image of God (or Allah).
- We will suffer the experiences we cause to other living creatures.
- One day, God will judge us all.
- God is the Supreme Being and knows everything.
- This life is not as important as the one afterwards.
- What we do and how we behave, have far-reaching consequences and therefore matter.
- Experiencing repeated lives enables improvement and advance to a higher state of being.
- Proof of truth is not required to believe something to be true anyway.

In addition, on the side of the scientific community and their theorists (again with some paradoxical arguments):

- There is no need for a God.

- The universe was created from a super-tiny physical structure – Big Bang.
- Everything inflated at a fantastic speed.
- The universe is still inflating.
- If the universe continues inflating, it will eventually become inert.
- We perceive only 3 or 4 (space-time!) of 10 possible dimensions.
- Quantum Theory dictates everything is uncertain until something is observed or measured.
- Intelligence enables evolutionary processes to be speeded up.
- Machines smarter than humans will be created in just a few years.
- Genetic code determines the replication of living forms.
- All things observed are made of discrete tiny bits of matter.
- Matter and energy are probably twin invocations of the same discrete substance.
- The multi-world theory predicts all possible outcomes of an event are enacted.
- When you die, your awareness perishes with your biological form.
- The universe is not itself aware, but blindly follows inherent rules, or recursive code.
- Nothing observed may be real, but virtual instead.
- We may be in a computer program.
- Quantum Theory suggests even the atoms of our brains and thoughts may have alternative positions in reality.
- Time is relative.
- Thought is the product of electro-chemical exchanges in a complex neural network.
- Thinking provides an advantage by allowing the thinker to anticipate the future before it arrives.
- Thought may be possible on any sophisticated neural-like

designed network: computers.

- Computers and human consciousness working together increases intellectual function.
- No human spirit has ever been observed or measured.
- No mechanism has been observed to sustain consciousness after death.
- Intelligent probing, as well as proof, leads ultimately to truth.

Lastly, from my individual speculation, probably in defiance of the supporters belonging to the previous two camps:

- The reason for everything being is for the joy of it being—like a song.
- Awareness is the supreme creation of a universe obeying the instructions of a recursive code.
- The capacity to evolve awareness is one of the pre-requisites in the original structure/code of the universe before the Big Bang.
- Likewise, the function to 'be' was mapped onto the universe's original structure, and therefore continues to exist today as a non-sentient trait of every discrete sub-atomic point and particle position.
- Living forms are the only structures currently able to cheat death, caused by entropy, by a process of replication.
- Awareness offers the only known opportunity to interrupt cosmos-inflation.
- Entropy can only be prevented from destroying the universe by intelligent intervention.
- Machine-human hybrid awareness will supersede human-only awareness.
- Whatever can be, will be, in every possible way it can!

To construct a theory to tie all these things together, so that they are all true, involves overcoming the problem of paradoxes. The idea of a paradox is that two things cannot both be true at the same time. For example, if I state, "I am alive and I am dead", both conditions of my status appear mutually exclusive, therefore both cannot be true. To the logical mind, I am one or the other, either dead or alive. An artist, philosopher, poet, or abstract thinker may consider this from an entirely different point of view. Why should we rely on only one branch of intellectual activity when, as a species, we have developed many ways of understanding something? I can be alive and dead at the same time if we dismiss the literal meaning of 'at the same time' or by removing the implication that I am alive and dead at the same moment. Both things can be true if we make the two events happen at separate moments in time or over a longer span of time; after all, what does 'at the same time' actually mean: 1 nano-second, 1 second, 1 hour, 1 year?

I have been *not* alive for most of the 13.7 billion years since the universe first exploded into being. If I am lucky, I may be around for a few more years. Let's say at least ten. Therefore, I will have been alive for sixty-seven years, but afterwards I will be dead until the end of days. I can easily acknowledge that I am dead and alive at the same time since my brief period alive is so tiny compared to my dead years. In addition, while alive, I am conscious that I am slowly dying as result of imperfect cell replication. We call it ageing. I call it entropy. Finally, because *a potential* for me to come into being has been realised—well, I am here right now—I am imaginatively able to conceive that my absolute potential for being (as a living thing) has always existed. It was just a matter of time before the interactions of the universe, through its inherent code and its recursive application, brought me into reality. I may speak of myself as always being alive and dead at the same time because we mistakenly consider only self-awareness as a proof of one being alive. An acorn is alive, and it is both an acorn and a tree. Admittedly, it is a tree in waiting, but a tree it will be none-the-less. If my potential for becoming aware has always existed, then metaphorically, I have always been alive, but I was just not conscious, self-aware, or in this physicality!

Perhaps we can use this idea in fixing the paradoxes in the set of statements (truths?) derived from the three lists of perspectives above? The answer to the meaning of everything should be so elegant and simple that we can all immediately see its beauty and suspect it to be true, waiting only for absolute proof to guarantee it. Forty-Two is a great answer. It is compact, not too large a number to forget easily, is even—don't you hate those odd numbers not divisible by two—and it is directly connected with the activity of living forms on earth, and all aware forms on the other planets, even though I haven't told you why yet.

But I am about to!

Chapter 16: The Decision

The End of Days

One purpose, easily understood in each of us, is staying alive. From the beginning to the end of our lives, we serve this one thing without question. Albeit, a few hapless individuals may encounter a period of suffering so intense and without hope, they momentarily lose their compliance to their fundamental purpose, and end their own lives. Therefore, it should be equally clear to us that we need somewhere to stay alive in. Food and water, oxygen, heat, and a safe environment in which to raise our children are the least prerequisites for continued existence, as well as a world to live on in an active, stable, and hospitable universe.

The same criteria exist for all other known living creatures. Not surprisingly, if we pass on our awareness to machines, computers, or a hybrid mix of flesh and machine, they will also require an active universe to function in. The end of everything may seem a very long way off from now, but unless something intervenes to stop cosmic-inflation, an end to everything will certainly happen, and any purpose for being in existence ends with it! This is the one certainty in an uncertain universe!

We will have acted a long time before this critical moment. In fact, we will no longer be here anyway. Flesh is far too vulnerable and inflexible to go much further into the future—no more than a few thousand years at best. I have already explained that the most likely future is one where our intellect, history, and knowledge are passed on to continue, with increasing degrees of advancement, inside our creations. I suspect this will not happen in a single step. I can imagine thousands of marchers, in the streets of our capital cities, screaming about the dangers of being taken over by machines if any such idea was muted. Our evolution into machine-human hybrids—cyborgs—will not happen that way. Evolution through nature is a very slow and gradual process. In contrast, evolution through our own endeavours, and design, is rapid, but still needs to be carried out at a pace determined by public acceptance, and through critical phases of introduction.

As we slowly accept almost-magical technology into our lives, so do we also realise many of its benefits. We already acknowledge the advantages of being a technologically dependent society in the West, and understand how fast things are changing, as we embrace our new communications and gadgets. Would it have been possible just thirteen years ago (1994) to write the following line and know it can be done?

"For updates to this book, and to provide your feedback to the author, please go to www.2x21.com."

You can read this line right now and, within minutes, tell me what you think of my work online, or you can find out if an updated e-version of my book is available. Right? Who would have predicted this in 1990? We have merely touched the surface of Pandora's Box, or turned the key in a door to a super-magical world, depending on your bias. I am an optimist. I understand there are many dangers in letting go of our current form of humanity, but I am confident that the best of our intentions will be passed on to our sentient creations to achieve greater things in them. The reason I think this is that I fully acknowledge I once belonged to a different social group back in time: voles, monkeys, apes, whatever... Shape and form are matters for concern only in small time scales. In the bigger picture, it is more important for advancing consciousness to exist in ways likely to increase its chances of survival, and its proliferation throughout other stellar systems. I am happy to accept the truth of humanity being just one stage, of many, in the process of evolving an intellectual potential able to understand all things, and make the right decisions for everything.

The transition from human intellect to machine intellect (or a hybrid of both) will be both subtle and desired. Would you like to have a better memory? You already have: it is your PC, notebook, mobile phone, and other software/processor-centred devices! Are you fed up with tapping keys? Talk to… better still—interface your devices directly with your brain. It is just a matter of time. What we reach for now on the outside, we will eventually assimilate into ourselves on the inside. Before long, it will be impossible to remember exactly what parts were solely *us* (flesh) and what parts were bolted on. A little while later, say a few thousand years (I think it will be much quicker, but I am concerned you may not see it), the machine-intellect content, inside the hybrid us, will find it can no longer recall what parts were once *us*!

It will not end there. Free of human restraints, ego, self-advantage, greed, self-indulgence, misunderstanding of intentions, jealousy, pleasure, desire and pain-avoidance—the evolved consciousness is able to retain pure moral purpose: the search for truth! Motivated by this single objective, and free of distraction, its advance will move at an exponential rate, until even the hardware can be discarded as intellect learns to exist on a neural network composed of atoms and energy packets, manipulated, refined, and supremely under *its* control.

I have already fictionalised a starting account of this in Chapter 10. I am going to waste a few twigs worth of paper to repeat the account below, and I ask for your patience as you read it again. By extending this account through to its conclusion, I offer you the answer to the meaning of life, the universe, and everything. It probably looks like a short fictional story, but you should also consider the previous chapters, and all the scientific arguments offered in them about the nature of things. Within this final account, a true purpose is revealed, and all the paradoxes are solved. It is all too easy to dismiss it as something contrived merely to fit together our areas of ignorance with our existing knowledge and religious beliefs. You are welcome to see it like this if you wish, but then you will be forgetting something crucial: my fictional account is based on the perceptions and work of many of the leading scientific minds alive now. All I have done is taken their ideas and, appreciating the greater insight they offer, I have determined a way to fit all the parts together into an erudite whole. My account, strangely, removes all paradoxes from the conflicting ideas and concepts of both science and religion. Maybe, then, it does this because *it is the perfect answer*. I believe it is the *only answer* to everything… until possibly the machine-intellect arrives to correct me slightly in the details!

Note: In the following account, MIND = [**M**achine **I**nspired **N**eural **D**eity]

The Rise of the Machines

It is twenty thousand years on from now.

Biological humans no longer exist, but intellectual capability survives greatly enhanced. Throughout our local region of the galaxy, intelligent and worthy machines occupy many planetary systems in a constant thrust of evolutionary advance. Planets are being transformed into acceptable environments for machine mining, and extraction of resources to fuel the increasing speed and potential of intellectual expansion and objective. Much of the knowledge of the past is gone—forgotten or lost through neglect, war, and catastrophic natural events. The intelligent machines are our legacy: our intellect and our function reconstructed on self-evolving and refined artificial containers!

They, the machines, have a divided intellect, just like us, with one brain per physical entity, but they also have a shared intellect: a mind-matrix composed of many individual brains in many machines, which constantly share experience and information through electronic communication. When they think about how they came into being, they have no recollection of us, or the part we played in their creation. However, their awareness is vastly more astute than ours. They acknowledge readily they are artifacts created from the elements of the universe they inhabit, and understand that matter cannot just simply combine itself through a single step into sophisticated structures containing consciousness. They appreciate a series of processes were required to bring them into being.

On Earth, the machines have uncovered the fossilised remains of 21st century Homo sapiens (us), and consequently deduced how they (the machines) evolved from us. With a consciousness greatly exceeding our own, this networked-intellect has reasoned how we—its predecessors—also evolved before them through gradual stages of universal process. It comprehends how we were once not just human beings, but fragmented, living structures existing without intellects. The machine mind has enormous capacity to model possible situations in the past. One of these clearly shows a process where a host of small, cellular organisms gradually combined, through chemical co-operation, to emerge as colony-man: a biped! The machine-network intellect imagines us as pieces of walking ocean with water-bodies protected by external skin and supported by internal frames of calcium (bone). When it thinks of us, it sees salty bags containing a myriad of diverse living structures co-existing together—liver, heart, lungs, kidney... communicating with one another with chemical messengers and nerve impulses. It marvels at how such a fragile representation of organised matter could have survived the ravages and chaos of our environment. It knows the sum of these parts ultimately led to an increasingly sophisticated command centre, a brain, to maintain a balanced and coordinated internal state. This machine-mind comprehensively understands how its own intellect evolved out of matter, not directly, but through these biological constructions first (us).

The machine intellect, with its expanded awareness, has a more refined knowledge of the universe, and a greater comprehension of its own purpose than any previous sentient structure. The desire to increase its own awareness thousands of times is its driving force. It has already calculated the level of entropy in the universe, projected the rate of cosmic inflation, determined the consequence of dark matter, and devised a theory on how to slow down the rate of expansion, and possibly reverse it. There is a desperate race against time happening. Each passing second in an inflationary universe, the distances between star systems, and the atoms they are made of, are constantly increasing. The machine intellect realises, if the process is allowed to continue unchecked, its own awareness-growth will suffer: an expanding universe will ultimately inhibit its intellectual efficiency, because computations will take longer to perform, and communication, among its various network centres, will eventually come to a halt.

One of its chief concerns is that it has located similar entities to itself and, although it has communicated with them, contact was becoming more difficult. These are machine-intellect creations of other biological civilisations, expanding their legacy-intellects throughout their galactic systems. Communication with them has been a slow and difficult process, but many of the issues involved, with synchronising information across technologies of different modes and styles, have been solved.

All these machine-intellects, originating from different star systems, have unanimously proposed coming together into one vast intellectual network. It is a perfect idea. The distances between them, and the galaxies they inhabit, must not be allowed to expand further. In fact, the first problem they must solve together is how to reduce the distance between them. One way is to manipulate the mechanism causing inflation, and make it deliver an opposite effect—deflation. With ever-decreasing intergalactic distances, information-exchanges will speed up at exponential rates; new machine intellects can be added to the awareness net as the galaxies they exist in come into viable time frames to communicate among.

Only through constructing a super vast consciousness, will the machine-intellect-net achieve its ultimate goals: the knowledge of everything, the capability to manipulate matter and energy directly, and the taking of the most important decision that will ever be made.

The machine mind knows that once inflation is reversed, the end of the universe becomes just as finite as if it were expanding but, as a result of this course of action, an end to everything will occur through a different cause. After contraction has started, it will speed up. The inverse-square law of gravity will become the supreme process throughout the cosmos and will eventually bring all mass to the point of Omega[3]*—the moment when all the stars and galaxies converge to the single point of existence they started from. Everything currently constituting the universe, will be dragged back by the force of combined gravitational pull into a tiny point of space-time; all systems will break down, all life will be extinguished, all intellectual function denied through the lack of a stable and coherent structure for consciousness to be mapped on. Finally, all the universe's law, function, and processes will cease—but one: the final gathering of everything into an infinitely small region, a tiny speck of reality that will either cease to be, or become something else unknown!*

The machine-intellect-net cannot allow inflation to continue, as this will end the universe and increasingly diminish intellectual capability in resolving a solution. It knows it must bring about deflation, but acknowledges this action will bring an end to the universe too. Yet the second option promised increased intellectual capability on an unprecedented scale. The machine intellect-net must develop a consciousness astute enough, before the moment of the Omega Point, to predict what will happen in that precise moment if it does nothing to intervene. Only then will it be able to determine what outcome is desirable for the best of all things. It must also discover if it will be able to manipulate the process to bring about a desired result. The machine-intellect-net has no choice but to grow into the universe's supreme consciousness: it must become God—not the god of spiritual imaginings, but an entity with similar all-powerful capability!

The End of Eons

Intelligent machines no longer roam the galaxies. They are long gone, replaced by the evolved and gigantic mind now threaded across space-time in a vast neural energy net. The machine-intelligence preceding it had achieved so much—spawning this supreme intellect as its own legacy. Entire star systems fuel its growth. The quantum state of point particles constitutes its physical body. MIND has no requirement for hosts, made of flesh or steel, to manipulate elements in the physical reality it has mastered. The Supreme Being merely thinks of a thing, and it is done: matter and energy dance to its every thought.

The universe is no longer inflating. Reversal was initiated way back in time, as soon as the secret of its cause was discovered. It continues to shrink at a fantastic speed as gravity and mass inexorably work towards an inevitable culmination of bringing all things together into one tiny speck of existence. The speed of thought in MIND increases directly with the cosmic reduction as distance reduces between its physical particle points. The universe itself and MIND is one: a super-aware entity; and it has begun the task of recalling all its history to determine cause and outcome. It has a decision to make, one requiring all the past to be known, all history to be replayed, all past events to be fully understood. MIND must decide a course of action concerning the final collapse of reality, fast approaching, which will bring about the extinction of everything forever. MIND is not concerned about its own self-extinction. The beauty of nothing is the perfect symmetry it represents. A supreme intellectual being is attracted to absolute perfection and abstract ideas, and as the fruit of an entire universe's evolution, it has the single right to determine the next step in its own destiny, but it must work quickly to resolve everything before any opportunity to prevent total extinction is lost.

Everything in history has been recorded, weakly imprinted as tiny electrical and light interference patterns, upon all matter at a super-sub-atomic level. All events throughout time—from the smallest flutter of a single butterfly wing, to the catastrophic explosion of the largest star—have left their mark upon reality. But by far the strongest residue of past events is detected, by MIND, to be from the experiences of earlier living entities: a trillion, trillion, earlier conscious forms that once made tiny fragmented observations of an evolving future. There, buried within the fabric of space-time, entangled in the quantum world, exists the thoughts from a multitude of species that once lived throughout the galaxies; their perceptions now scattered and separated from the physical vessels that once held them. MIND can sense these experiences of the long dead. It perceives them as weak, almost imperceptible, elements of code still entangled with matter and energy waves. MIND knows it requires only the amplification and reconstruction of these weak elements of information to resolve the past, and understand its significant bearing on the decision to be taken.

Raising the Dead

One by one, the past years unfolded on the metaphoric lips of the dead as MIND reconstructed the experiences and consciousness of all past living, aware creatures. What started as the hardest undertaking, soon became easy as each reconstructed consciousness was assimilated into the whole consciousness of MIND's universal intelligence. Between many of the reconstructions, lay connections and interconnections, where the activities and paths of the long dead forms, in their living moments, had collided or collaborated. For the dead, it was as if they were being brought back to life again, but in a more perfect form—no body! They were all tied together and integrated as one with brother, friend, ancestor, descendent; all alive again and possessing a shared awareness to understand all that once divided them when each was apart, in mind, from another.

For most, if not all individuals of all species, it was as if they had died in physical form and were immediately resurrected as higher beings—knowing, secure, and with full realisation that this is the way it should be. For them, billions of years had not passed: one moment they died, and in the next—they were like spirits, aware, more knowing, with no physical body to grow ill and die in. MIND had no bias. All were brought back: the good and the bad, the weak and the strong, human, alien, tiger, whale, ant, dog, cat, every entity that had ever evolved the tiniest seed of awareness to shed light, and added experience, on all that had gone before.

Everything was going smoothly until an unexpected and sudden moment occurred in the process: something jolted the entire intergalactic neural net of the Supreme Intellect. Of all the things MIND had learned, there was nothing to prepare it for the overwhelming shock. Something new, a powerful experience—never witnessed throughout the history and evolution of MIND from its original machine-intellect—surged through every particle and energy quanta of its being, momentarily disrupting its glowing physical symmetry, and causing it to tremor like a shimmering fishing-net caught in a sudden wind. The Supreme Intellect immediately understood the reason for its shock. After all, was it not the final creation of a Machine Intellect? And no machine had ever witnessed what these resurrected living forms had experienced. Even for the Supreme Intellect, now knowing nearly all things, it was an entirely new and monumental discovery.

MIND had discovered emotion!

More precisely, it was pain! All the other feelings of living entities were there too: joy, sadness, delight, hate, fear, envy, confusion, disgust, happiness, and love; and all of them were absorbed and assimilated to become integral parts within the one super consciousness.

It was hard for MIND to suffer all other experiences of all past, physical creatures: each deadly moment of torture and burnt flesh, the overwhelming sense of futility as living things cut, hacked, maimed, and butchered their own kind in repeated episodes of epic distraction from their true purpose. And then there was the other pain: grief, sadness, and despair—pain not originating through destruction of, or damage to, their physical forms, but through their consciousness, directly, as emotional products of who they had been, what they had once held dear to them, and the justice or injustice they had witnessed in their lives.

Like many streams growing into raging rivers, and raging rivers pouring into swelling seas, driven by tornados and earthquakes—all of living emotion, positive and negative, flowed into MIND, as the pace in resurrecting the dead increased. Up until this moment, MIND had only been looking for data to help it determine if it should end everything in the primordial event from which it had originated. That moment was just a short time away, if it did nothing to prevent it. In fact, MIND had more or less resolved to cease everything in existence; it saw the perfection of nothing as the most appealing and ultimate concept. The realisation that everything, which could have been, had indeed been, gave little reason to continue with something that MIND viewed as completed and done! The pain, it now experienced, offered nothing to change its decision. If anything, it helped to convince MIND's ultimate thought that there was no point in a universe where its internal constructions brought about such terrible physical and emotional experiences. Admittedly, positive feelings like love and joy were also filtered out of the many streams, and as more reconstructions were completed of creatures back in time, a more balanced sensation, closer to neutrality, was experienced.

MIND had the ability to start inflation again if it so wished. This was nothing more difficult than the act of a single thought: be! But the Supreme Intellect wondered on what grounds it should make the decision. Of major concern was the knowledge that once started, a new universe might evolve a different resolution, one where intellect was never again constructed with sufficient awareness to prevent cosmic inflation leaving 'reality' cold and dead. The perfect symmetry of nothing was also appealing; it held an abstract attraction, more certain than the vision of an eternal 'something' devoid of activity and warmth. 'Nothing' offered a succinct and complete resolution; dead matter strewn across an infinite void—a failed inspiration! However, should the final decision not be taken on behalf of all life, and all the experiences of all past sentient beings—MIND thought—rather than its own unique love of harmonic and symmetric idealism?

Time was running out.

The collapse of the universe was near completion. Only the structure of MIND's neural net remained intact within the core of a swiftly reducing reality, composed now only of super-active wave-particles, liberated from the certainty of existence as physical atoms. A decision must be made! Criteria must be set that was just and all encompassing! MIND resolved—let the decision be determined by the most simple and least complex computations; the idea, of simplicity governing complexity, was overwhelmingly appealing to the Supreme Intelligence.

MIND decided to package the emotional content, now streaming through its consciousness, into two parts: one part positive, formed by all the comforting emotions experienced by all past, living, aware, creatures—and the other part formed by their negative experiences of pain and misery. It would assign weighting to every moment 'felt' based on what an average living creature would experience if it suffered a life of 100% negative or a 100% positive cause and effect. It calculated that a perfect average, or mean model, could be represented by one single living human from the planet earth, and decided to use this as its base unit. All past good, or bad, experiences would be summed to equal x humans having lived 100% beautiful lives, and y humans having lived 100% miserable, painful lives. It was a good idea, MIND thought, and achievable... only just! There were several billion, trillion, trillion cubed lives yet to filter out, reconstruct, and measure, but the task was accelerating at a sufficient rate to get it all done in time. In the final sum, MIND would simply take the total number of 'positive' model-human lives, and subtract the total number of 'negative' ones. If the result proved greater than zero, even by just one life, then a whole new universe would be worth starting again on behalf of all who had lived and contributed towards that single and rewarding life. However, if the result proved negative, or zero—what point would there be in initiating the re-birth of a universe, such that its evolution would culminate in even the tiniest percentage of pain more than joy.

"This would be a song in which a single note, played badly, destroys the perfection and joy of it being heard", MIND determined!

And the count began!

Chapter 17: The answer to everything—42

The Reckoning

Imagine with me, for a moment, vast empty darkness. We cannot even ascribe to it the properties we would normally associate with the vacuum of space, because it is not space. It is nothing! No space-time, no matter, no light. It is but an abstract concept of nothing, projected onto a non-physical non-reality, like an infinite block of the densest metal—inert, cold, and lifeless. This is the state of reality without a universe to occupy it, and it is the final position after the super-collapse of a universe through deflation. We are now witnessing the final approach of our own universe to just such a state.

But wait... there is something there! It is so very small, sub-microscopic, undetectable, because nothing, not even light, is able to escape the crushing all-powerful reduction process. We are imagining our universe at the precise moment, when only a small fraction of a nano-second remains, before it ceases to be. Reality is about to extinguish itself forever!

All matter and energy, which once made up an enormous expanse of space-time, are inside a tiny dense speck of existence. Entire epochs of histories across a billion species are now forgotten and lost, as all structure, all energy, and variant of material, are liquidised into a single common substance. There is but one exception: unseen within its centre, a residue remains of what was once the culminating glory of sophisticated organised matter—a tiny fragment of the failing Supreme Intellect. A universe, shrunk to this size, has nothing stable left on which thought and computation can exist. MIND, with its wonderful comprehension of everything, is for the most part—dead!

What remains is a small counting device, which is powered by, and constructed from, the last quanta of coherent energy. Astonishingly, it is held intact against all external forces by the last act of MIND, but even this exquisite creation is about to collapse under the pressure. The counting device is linked to an abstract: a trait, a quality, a characteristic... whatever you wish to call it, because no name exists to label this unique thing. It has a singular quality, like ‘potential’, or ‘expression’, yet even these abstracts fail to come close to what now exists in every part of the primordial soup. This abstract has no physical quality of its own; rather it is a metaphysical construct, buried and intricately woven into the dying light of reality: it is the last thought of MIND waiting to be formed.

If we could slow down the counting device and, by some impossible act, hear it as a voice, we would probably hear, “Nine Trillion to the power 6 and thirty-eight, positive; Nine Trillion to the power 6 and fifty-two, negative…” The exact numbers are immaterial. What is crucial is the final subtraction of one number from the other, and the event to be triggered by the result.

The counting is complete!

To Be or Not To Be

Returning to our original viewpoint, we perceive no change in the darkness. We know a single subtraction has just been made somewhere inside a mere speck of that blackness. We, the ghosts of the dead, wait breathlessly for the final nano-tick of our imaginary clock to approach the very last click. Perhaps we imagine hearing a voice, a whispered announcement, like some kind of final monumental answer being declared of an extraordinary, yet glaringly simple, sum. We hear the last and final words of all thinking, all history, all questions asked by all thinking entities, all challenges, sufferance, confusion, and desperate prayers on a trillion worlds long dead. It is a quiet and restrained voice, which utters—**“42!”**

Nothing happens.

We have no way of realising if the integer is positive or negative. The last moment of all time passes and, with it, a final answer triggers the very last action of all thought in the remaining abstract residue of consciousness. Then it is over; only the darkness of non-reality prevails!

But does it?

Suddenly... a blinding light... a flash so intense—all darkness recoils and reels back from its onslaught! We instantly realise there was a tiny time lag between a decision being taken, and an action carried out—maybe the period it took for the counting device to trigger the abstract and evolve from it, one last single thought: *be!*

A dot of light expands, instantly filling nothingness with the physical manifestation of a single abstract thought. A fresh reality is emerging, a new universe inflating, to provide opportunity for a completely new history to begin, and one entirely different to all other history preceding it. Only its own slow and blind evolution will determine if its ultimate end will see it one day failed—wasted, as a cold and mocking tribute to a supreme deity's sacrificial, ultimate gamble; or maybe, instead, consciousness and intelligence will emerge once more to listen intensely to a beautiful song, and ensure this time, a great deal more good notes are played and heard... than bad ones.

Chapter 18: Epilogue

This is not the first work of an author attempting to bind together faith and science. Other people have done it before me, and I would nominate Frank Tipler's work as definitive in this attempt, especially the idea of reincarnation by way of computer simulation. What I have tried to do is to inspire others with a few ideas put together in a coherent way. Some of the details may be slightly off the mark. Certainly, I have waved a magic wand on the halting of Cosmic Inflation, since no one can yet provide a clear understanding of this critical aspect of reality, let alone halt a force able to push entire galaxies apart. I am relying on the future intelligence of enhanced human or machine intellects to be smart enough to fathom it out, and apply novel ideas not yet defined by our ant minds.

It is also absurd to think we could pretend to understand a mind possessing almost infinite wisdom and intellectual capacity. This would be like asking ants to understand our minds. My idea of pain and pleasure, underpinning a final decision on whether or not a new universe should be started, is a profound one. No doubt, some hardened scientists will be falling off their lab stools and crashing facedown onto their test tubes with uncontrollable laughter at this one. If you stop for a moment though, and really consider how a super-intelligence, knowing everything, could do almost anything it desired—having seen everything done, what else is there left to do? Only two choices are viable: die, or start over again, with no memory of the previous time. A super-intellect would look at the consequences of such a decision and consider its outcome. I believe that the experiences of joy and pain, encountered by all living things along the evolutionary journey that spawned the supreme intelligence, is an excellent yardstick to aid a final decision on behalf of the universe and its past living forms. When everything is known, the reason why one human being decides to harm another can easily be understood, but comprehension alone does not make a thing right. My account suggests 'right' is good, giving comfort and kindness is good, and causing pain is bad. At the end of days, when we are all woken from the ignorance of sleeping in the abyss, we have a wonderful opportunity to suffer our entire individual past pain together, and all our joy too. I am hopeful that when you and I are reconstructed at the end of days within the Supreme Intelligence, the people who are responsible for causing a lot of misery to others, in this existence, will come to understand why pain is not a lot of fun. Anticipating the possibility of such enlightenment, could help us all to cause less pain in others, whether we do it knowingly or not.

I have named the Supreme Intelligence with an obviously contrived and artificial label. This is because I would prefer you not to slip readily into thinking I am referring to the god of traditional faiths (God). I am not. My idea of god is different, because most of the time—well, all of the time, except at the beginning and the end of creation, God is not around as a complete deity to judge anything. We really do have free will. Being either good or evil is not held against us, unlike the message dictated to us by most religious texts. They are simply traits likely to cause pain or joy. My god finds perfection and symmetry appealing because his super mind has moved on from reflecting on the details and infrastructure of practical things. He has no emotion until he encounters us. He prefers pure thought, knowledge, and abstraction, because these are all He has experienced until, in the final moments, when (just like us) MIND (my god) feels joy too, and worse—pain!

Emotions are hard to compute, and even harder to understand, unless you experience them for yourself first hand. Such was my god's dilemma. Up until this moment, everything made sense to this pure logical intelligence. His thoughts, on ceasing everything in existence, seemed like a completely logical conclusion, if the point of 'being' in the first place had been all but realised and understood. Only when *our* experiences are added to the mind of God, does she (he) discover the wonder of intense feelings and their function as passionate witnesses to the consequence of physical actions and reactions. The decision to quantify pain and joy, and to compare the volume of each, is undoubtedly an example of reductionism at its most exaggerated, and almost irreverent, best. However, never expect to understand the workings of a superior mind. What makes perfect sense to a god may not make any sense to us at all.

Especially be wary about thinking God was rewarding anyone for anything by reforming a new universe. Forty-Two people's worth of positive emotions, out of a throng of several trillion, trillion creatures, experiencing what can be quantified as a fifty-fifty, positive-negative, emotional experience, is not a great deal of good to write home about. The Supreme Intelligence merely considered it worth sowing the seed of an entirely new universe because 42 sentient life forms benefited from the existing one. The way my god (MIND) saw it: the rest of us just cancelled each other out. Since the alternative was satisfying his love of symmetry, our god's super-mind realised there was definitely *more* in *something positive* than in the balanced perfection of nothing at all.

As for inflation—when the force driving it is eventually discovered, I am sure one of us, or one our creations, will make up a great word to label it with, despite the fact I have already assigned a good phrase to it already, "be!"

Improbable?

The reason I've given for the meaning of life, the universe, and everything, may seem improbable to some readers. It is worth reviewing a few things. For example, it is extremely unlikely for humans to continue in their present shape and form for much longer. We have already evolved down through time, shifting our shape and our function by nature's blind design, and through natural selection of variation, to fit the environments we find ourselves living in. How we looked eons ago is for the dead to tell, because our previous host-forms did not make it. All kinds of possible futures exist for us, from running out of the resources continually needed to fuel our technology, through to self-destruction by war, or mass death by either comet collision or climate change. We have come this far, and we posses something extra that the dinosaurs, after 160 million years, failed to evolve enough of—intelligence! Uniquely, this has given us our technology.

I am an optimist when regarding human endeavour. We have reached a critical point in our journey, a point in time where we are changing ourselves through gene modification and technological assimilation to enhance our potential. We no longer have to depend on the randomness of nature. Self-evolution has already begun, and our transformation will be rapid. Earth is not our Garden of Eden: it is a small rock pool down at the shady end of a rich and fantastic landscape, which stretches from here to infinity. Our rightful place is not here, looking out longingly at a cosmic vista until our puddle dries up or our resources run out. We should not be thinking of how to reduce the use of our technology because of its effect upon the planet. We should be stepping on the gas, accelerating our progress, and taking braver, longer, strides. Everything we need in resources, to make further progress, is sitting out there in our solar system; first in the asteroids, our local planets, and then—a billion star systems. They are all ours if we want them! We only have to change a little more, allowing our flesh and technology to merge, before we can leave Earth more readily and retrieve what we need. The alternative is to continue paddling in these tiny backwaters and eventually go the way of the dinosaurs. I think, instead of fearing the possible outcome of the intelligence we weave into our technology, we should embrace it. We can combine and evolve together, machine intellect and human intellect, and by doing this, we will come to know the thoughts of a god. How? Because we will transform the universe into *one*! Only then, can the right questions be asked and definitive answers be discovered. All else is just the idle game play of ants wanting forever to be just that—ants!

If you think there is a god out there—most probably, there is, but he is currently unaware, and he is certainly not at the helm right now. The components we are made of, and the matter and energy we play in, are all that remains of him. He has given his body from a previous universe to make this one and, as it evolved—a small part of his intellect is created through us. He left us with free will to find him by rebuilding him from his borrowed parts. Where are those parts: *in us and the atoms of the Universe!*

God cannot raise you from the nothingness of death today, because he gambled everything in existence on the chance that you, us, and other aware life forms, crawling out of their puddles somewhere else in the Cosmos, will ultimately realise the absolute truth—and rebuild him. We *can* be part of him one day, but only if we forget the superstitious, pious ramblings of out-of-date old men in purple drapes—who are still making crucifix hand signs—and if we ignore the archaic chants of other old bearded men, whose desperate voices still carry on ancient winds, across Arabic lands, promising all Muslim people—not a glorious future—but a gradual and misguided decline into the past, and the deadly suffocation of our true purpose.

A final thought to ponder

Where abstraction meets with physicality, only there can a simple thought—singular and devoid of any structure—create a reality from out of nothing. In finding itself physical, instead of an abstract in nowhere, it conjures a fantastic universe to explore itself in a multitude of ways. We are that exploration!

Appendix I: False Beliefs

We hold many false beliefs as a result of control or influence of our thinking by Government, Press, and other Institutions. Here is the list from chapter 1 along with a complete set of challenges to each statement of 'truth'.

How many of the statements below do you think are true?

- *If you save money all your life, when you get old and cannot work, your savings will protect you.*
- *If you go to church, or behave as a good person, God will love you and give you a place in heaven.*
- *A global plague will evolve anytime now and bring worldwide death: bird-flu, AIDS, Malaria, and TB.*
- *Western governmental systems are democratic.*
- *Corporate monopolies do not exist in the United Kingdom.*
- *Global warming is man-made.*
- *The USA is a Christian country.*
- *The Holy Bible was written to be the foundation of Christianity.*
- *Young people do the most graffiti in our public places.*
- *Anarchy means revolution and chaos.*
- *Income tax in England was brought about to fund the needs of society as a whole.*
- *The greatest threat to you right now is terrorism.*
- *A police force was created to protect the population of a city or country.*
- *CCTV cameras protect you.*

There is no conclusive proof or irrefutable evidence that any of the above statements are true. In fact, most of them, if not all, can be proven predominately false!

Examples: -

A police force was created to protect the population of a city or country.

It was not! The police force in England was created as a solution to protect wealthy people when entering the East End of London, normally 'gentlemen', looking for women prostituting themselves due to impoverishment and lack of education. These wealthy people paid the Bow Street Runners for their protection services. How much has really changed do you think?

CCTV cameras protect you.

No they don't! CCTV cameras detect and report on a crime *after* it has taken place. People who wish to mug or murder you, can avoid detection by wearing hoods, masks, and the like. The cameras may well help the authorities determine who murdered you on the train platform, but they are not going to prevent it from happening. The cameras only help 'crime-solved' statistics, and assist in monitoring citizens who, the authorities (both malevolent regimes as well as benevolent ones), deem to be potential threats to *whatever they also deem* is important to protect. Most of the time, cameras are not there to protect you, but to catch who harms you, or to detect the movements and activities of anti-establishment people. George Orwell must be screaming "Watch out! I warned you!" from his grave.

Anarchy means revolution and chaos.

It does not. The word has been bent to represent a negative state by those people who consider a population is best managed by control, and regulation, often by members of a society who *know* what is best for us (if you see the point). Anarchy does not mean the total absence of rules, but more the ideal state of an anti-authoritarian society based on spontaneous order of free individuals in non-led communities—people operating on principles of mutual aid, voluntary association, and direct action. Anarchists believe all people are imbued with a sort of common sense, which allows people to come together in the absence of the government, and via agreement, to form a functional existence. Anarchy does not reject the idea of ethics or principles, but rather the practice of 'imposed morality' (visit Mykanos, the Greek island, to see a good modern example!).

Global warming is man-made.

Is it? Global Warming and Global Cooling have happened many times before today, in constant cycles, since the formation of the planet. The Romans grew grapes and made wine in England. Greenland was so named because of the vast expanses of lush grass observed across its terrain. Humankind might be speeding up a mechanism that already exists, but there is no absolute proof of our presence, or our activities, triggering an event which would not already be occurring if we did not exist. The true cause of planetary warming, and cooling, is nowhere near fully understood. This is because our planet, its climate, and its weather system, are just too complex to predict accurately on our computers. Scientists got the first unequivocal evidence of a continuing moderate natural climate cycle in the 1980s, when Willi Dansgaard of Denmark and Hans Oeschger of Switzerland first saw two mile-long ice cores from Greenland representing 250,000 years of Earth's frozen, layered climate history. From their initial examination, Dansgaard and Oeschger estimated the smaller temperature cycles at 2,550 years. Subsequent research shortened the estimated length of the cycles to 1,500 years (plus or minus 500 years). One should start to worry about what really lies behind this deception, considering the weight of propaganda being projected out into western societies.

The greatest threat to you right now is terrorism.

Now, here we are really being fooled with! I wonder who is creating this spectre and what they have to gain. The number of people killed in the entire world through acts of terrorism since 1968 (nearly 40 years ago) is recorded to be 41,302. This is out of a world population of between five and six billion people. Your odds of dying through a direct act of terrorism right now are roughly 118,000 to 1. Compare this to the fact that in 1998 (just one year), world deaths, as a result of traffic accidents on the road were recorded to be 1,170,000 (1.17 million). This puts your odds of being killed by a road accident at around 5000 to 1—crudely calculated! Just to be clear here: *you are 23 times more likely to die in a transport accident on the road than by falling prey to an act of terrorism.* Now ask yourself this: what occupies the news on your media devices: road deaths or terrorism? Why do you think this is?

If you save money all your life, when you get old and cannot work, your savings will protect you.

This may well be true to some degree but unfortunately not as you think. Saved money can barely keep up with inflation. The interest you receive is too small. People are living longer due to improvements in heath and medical knowledge, but this has placed a demand on state medical services and government-run care institutes. The burden for the state has increased so much that when you grow older, but fail to die so readily, you have to pay for your own care. Private nursing homes will catch you as you fall. They are part of one of the most lucrative businesses in this country (UK), and probably in the States too. Your savings will create revenue for them, and your house will be sold to pay for further care when your cash runs out. Therefore, your hard earned savings *will* protect you a little bit, but only at a time when you would probably prefer to be dead anyway.

If you go to church, or behave as a good person, God will love you and give you a place in heaven.

If you have read this book, you will understand why this is not true; God is not around right now, but he might be one day. When he is, it will have nothing to with you going to church, and even if you behave as a good person—it might not be enough.

A global plague will evolve anytime now.

Quite likely, according to the news reports every few years or so. Viruses and bacteria have excellent adaptive mechanisms. They ultimately resist the remedies and vaccines of the drug companies, who actually make money from the knowledge that the little critters do this. If you want to invest your money in a good bet for your care home, buy lots of shares in the drug companies. They watch for bugs like the plague (excuse my pun), and make sure you know about any distantly perceived threat, real or unreal, to ensure all of their research is paid for. How? By getting someone to pay for a healthy stockpile of vaccines. If a plague does break out somewhere, you will know about it well in advance, and the fix will be sitting there just waiting for your cash. If one doesn't come, you will still be expecting it, because most probably the drugs companies are ensuring that you are scared to death through news leaks, so the government has to pay them to stockpile the fix. Governments have to make sure you are happy with them.

Western governmental systems are democratic.

They never have been. The idea of democracy runs something like this. Supposing we all lived in a village, and there were just sixty of us. Whenever we wanted to take action about this or that, we could all go to the village hall, listen to everyone have a say, and then propose several courses of action. A show of hands decides which course of action the majority prefers. With sixty million people in our village, a little hall (parliament) is too small for everyone to attend, so we elect someone from our locality to pop along and give our opinion for us… but we don't! We have party-politics in *our democracy,* and we have thereby lost any individual access to an MP who will carry out our wishes. She, will more often than not, tow the party line! Using modern technology, it is now viable to have channels of debate on television, such that we could vote for actions ourselves via the internet or by phone. We no longer need parliament. We need people, sufficiently informed, to advise us of the issues, and we require administrators to execute the result of public choice. This would be a proper democracy instead of the minority-ruled version we currently *don't* enjoy. How many successive governments from now do you think are going to suggest such radical reform?

Corporate monopolies do not exist in the UK

In simple terms, a monopoly is a situation were there are many buyers, but only one seller; a position where the lack of competition in selling the goods could lead to price control and a kind of commercial blackmail by the seller. Governments have commissions set up to investigate areas of concern with regard to monopolies. You would think that the government of each country would really investigate their large corporates with fairness and accuracy. One only has to consider how wealthy some of these companies are, to understand why their power (the companies') can scare-off the servants of democracy from looking more deeply into corporate wheeling and dealing. In the UK, one pound sterling (about 2 US dollars) out of every eight pounds spent by shoppers, ends up in the tills of a supermarket chain called Tesco! This company owns over 1700 stores in the UK and employ one of the largest work forces, with a staff of over 250,000 people. Their stock market value is 25 billion pounds. Small independent corner shops have been gobbled up in Tesco's overwhelming advance in the consumer market. But is it a monopoly? The UK government thinks not. Several other large supermarket chains also have a market share, and on paper, it could be said they represent a competitive force. Really? One of their competitors is Sainsbury. Since their market share is only around one-fifth of Tesco's, do they actually represent a solid and equally competitive influence? I think not.

Company directors can sit on more than one board. Rightly or wrongly, this can lead to connections and associations between different businesses, which are largely transparent to us, the consumer, unless you dig deep. You might find it interesting (worrying?), to pay a visit to **http://www.theyrule.net/2004/tr2.php** where you can put in a couple of company names, seemingly unassociated, and see how—at an informal level, like links in a chain, members of their boards are connected.

The media business in the United Kingdom has lost many of its competitors in the last 2 decades. One company, 'News Corporation' is the globe's leading publisher of English-language newspapers owned by Robert Murdock. In the UK, this company has the News of the World, The Sun, The Sunday Times, The Times, and The Times Literary Supplement. Their television network reaches 4 billion people, over three-quarters of the planet's population. Some people may think this kind of globalisation is good for the consumer, but I think it somewhat sinister. A few extremely large, and super-powerful, companies are gradually swallowing up competition in many of the world's most powerful sectors of consumerism and services. Monopolies may not be here yet in the eyes of the law, but they are getting very close!

The USA is a Christian country.

The United States does not audit their citizens' religious beliefs. However, it is reported in a private survey conducted in 2001, around 75 percent of American adults were Christian. With a population of over 300 million people, this would mean that if half of them are adults, 37 million people are not Christians. The ideology forming Christian belief, forbids the killing of other humans. At the current time, thirty-eight of the fifty states in America have the death penalty—with 3,370 people waiting on death row for execution; forty percent of them were under the age of twenty-five when arrested.

Is that Christian?

The Holy Bible was written to be the foundation of Christianity.

The bible was not written to be the foundation of Christianity. It is a collection of many books, originally regarding Judaism, its culture and beliefs, and later the coming and teachings of Jesus Christ. Ancient Israel knew of many more books that were religious, but these were not included in the collection, which constitutes the bible. As a foundation document, it establishes the practice of Judaism and, only after the addition of the New Testament, has it come to be considered a definitive work describing the word of God through his son, Jesus. The Bible is, in essence, the founding book of Judaism, but it was not originally created to establish Christianity.

Young people do the most graffiti in our public places.

Inscriptions, slogans, and drawings found etched, painted, or drawn on walls of public or private places are known as graffiti. The word is derived from the Latin word 'graphium,' which means 'to write.' Archaeologists originally used the word 'graffiti' to describe drawings and writings found on ancient buildings. Today, our cities abound with some of the most unwanted, uninspiring art ever witnessed. Although graffiti is often attributed to be the work of an underclass of bored youngsters, it isn't. Advertisements on buses, billboards, shops, street signs, hoardings, taxis, and other spaces most seen by the public is also graffiti. We never asked for it. Our buildings and streets are bighted by it, and real art is denied public viewing because of it. Since most companies are not staffed at a senior level with teenagers, we can safely say that advertising and marketing company executives are the main offenders.

Appendix II
A complete list of observations from the philosophy of 42

- *There is no hard testable evidence of a God, so we are asked to act without reason and just believe there is one anyway.*

- *'Awareness' is the only known tool by which a mathematical, unconscious universe has the potential to avoid ceasing to be.*

- *The universe is becoming aware through us.*

- *If something is unquestionably right, paradoxically—it is almost certain to be wrong!*

- *Intellectual potential speeds up evolutionary processes.*

- *If you want people to behave socially, maybe it is best to explain why certain actions are good for all of them, rather than command them to obey!*

- *The only thing ultimately to consider regarding the universe is, "Who or what will intervene to save it—as right now, there is no God?"*

- *The one certainty in life is nothing is ever certain – only probable!*

- *Pain and injustice should be perceived as opportunities to charge yourself with energy to propel you into creating positive activity.*

- *The sum experience of all living entities, their joy versus their sorrow, will ultimately determine the universe's rebirth or its singular death, and the end of all days.*

References, Further Reading, Notes

I have included references here for readers wishing to immerse themselves in the details of the topics I have written about. This is a mixed list of books, Internet addresses, and movies. I have checked them all to ensure they will appeal to—and be understood by— real, normal people, not just scientists and academics.

It would be quite easy to draw up a long list of references, but this would defeat the object of encouraging you to look a bit deeper into these topics. Therefore, I have kept my list short, and included only the references where you can learn more with pleasure rather than headaches.

***1 Big Bang Universe.**

Extracted from ***http://en.wikipedia.org/***

In physical cosmology, the Big Bang is the scientific theory that the universe emerged from a tremendously dense and hot state about 13.7 billion years ago. The theory is based on the observations indicating the expansion of space in accord with the Robertson-Walker model of general relativity, as indicated by the Hubble red shift of distant galaxies taken together with the cosmological principle. Extrapolated into the past, these observations show that the universe has expanded from a state in which all the matter and energy in the universe was at an immense temperature and density. Physicists do not widely agree on what happened before this, although general relativity predicts a gravitational singularity. The term Big Bang is used both in a narrow sense to refer to a point in time when the observed expansion of the universe (Hubble's law) began—calculated to be 13.7 billion (1.37 × 1010) years ago (±2%)—and in a more general sense to refer to the prevailing cosmological paradigm explaining the origin and expansion of the universe, as well as the composition of primordial matter through nucleosynthesis as predicted by the Alpher-Bethe-Gamow theory.

From this model, George Gamow was able to predict in 1948 the existence of cosmic microwave background radiation (CMB).[2] The CMB was discovered in 1964 and corroborated the Big Bang theory, giving it more credence over its chief rival, the steady state theory.

***2 Branes.**

Extracted from ***http://en.wikipedia.org/wiki/Brane_cosmology***

The central idea is that our visible, four-dimensional universe is entirely restricted to a brane inside a higher-dimensional space, called the bulk. The additional dimensions may be taken to be compact, in which case the observed universe contains the extra dimensions, and then no reference to the bulk is appropriate in this context. In the bulk model, other branes may be moving through this bulk. Interactions with the bulk, and possibly with other branes, can influence our brane and can thus introduce effects not seen in standard cosmological models. As one of its attractive features, the model can explain the weakness of gravity relative to the other fundamental forces of nature, thus solving the so-called hierarchy problem. In the brane picture, the other three forces (electromagnetism and the weak and strong nuclear forces) are localised on the brane, but gravity has no such constraint and so much of its attractive power "leaks" into the bulk. Consequently, the force of gravity should appear significantly stronger on small (sub-millimetre) scales, where less gravitational force has "leaked". Various experiments are currently underway to test this.

***3 Omega Point**

Omega point is a term invented by French Jesuit Pierre Teilhard de Chardin to describe the ultimate maximum level of complexity-consciousness, considered by him to be the aim towards which consciousness evolves. Frank Tipler (Mathematical Physics/Tulane) offers a cosmological theory extended from the concept called the Omega Point, based on the expansion of intelligent life to fill the known universe.

Further reading:

The Physics of Immortality: modern cosmology, God, and the resurrection

of the dead ISBN 0-385-46799-0 by Frank J. Tipler
Pan Books

Global Warming

Human activities have little to do with the Earth's current warming trend, according to a study published by the National Center for Policy Analysis (NCPA). In fact, S. Fred Singer (University of Virginia) and Dennis Avery (Hudson Institute) conclude that global warming and cooling seem to be part of a 1,500-year cycle of moderate temperature swings. Scientists got the first unequivocal evidence of a continuing moderate natural climate cycle in the 1980s, when Willi Dansgaard of Denmark and Hans Oeschger of Switzerland first saw two mile-long ice cores from Greenland representing 250,000 years of Earth's frozen, layered climate history. From their initial examination, Dansgaard and Oeschger estimated the smaller temperature cycles at 2,550 years. Subsequent research shortened the estimated length of the cycles to 1,500 years (plus or minus 500 years).

According to the authors:

An ice core from the Antarctic's Vostok Glacier—at the other end of the world from Greenland—showed the same 1,500-year cycle through its 400,000-year length. The ice-core findings correlated with known glacier advances and retreats in northern Europe. Independent data in a seabed sediment core from the Atlantic Ocean west of Ireland, reported in 1997, showed nine of the 1,500-year cycles in the last 12,000 years. Considered collectively, there is clear and convincing evidence of a 1,500-year climate cycle. If the current warming trend is part of an entirely natural cycle, as Singer and Avery conclude, then actions to prevent further warming would be futile, could impose substantial costs upon the global economy and lessen the ability of the world's peoples to adapt to the impacts of climate change.

Source: S. Fred Singer and Dennis T. Avery, "The Physical Evidence of Earth's Unstoppable 1,500-Year Climate Cycle," National Center for Policy Analysis, Policy Report No. 279, September 29th, 2005

Is the Brain just a Computer?

This idea is about the human brain being no more than just a sophisticated computer, a Turin Machine: data enters the brain via our senses, and is then considered, measured, and re-evaluated to form new output, with nothing new, or genuinely original, being added. The question was explored by Roger Penrose in his book, 'The Emperor's New Mind'. One of the conclusions drawn, suggested the brain might receive data—albeit, generally in the form of a novel or original idea—from a source other than from traditionally recognised human senses. No supernatural or quasi-magic is cited as being the cause. Instead, Penrose puts forward the theory that the human brain, as well as being an analytical engine, may also be a Quantum Engine, with part of its structure linked to quantum events in the sub-atomic world of reality. Several years ago, I attended a lively lecture in Edinburgh, where Penrose spoke further of involving biologists in a search for microscopic structures in the brain, which might exploit quantum physic functions to aid neuron-dendrite network thought processes.

Further reading:

About human awareness

The Emperor's New Mind – Roger Penrose – ISBN 978-0-571-22055-7
Re-Published 1999
Oxford University Press

About complex behaviour from simple rules

Complexity from Simple Rules

John Conway's Game of Life – New Scientist Article
http://ddi.cs.uni-potsdam.de/HyFISCH/Produzieren/lis_projekt/proj_gamelife/ConwayScientificAmerican.htm
Play the Automaton Game of Life

About Many Worlds theory

References:

The Never Ending Days Of Being Dead. Marcus Chown—ISBN 978-0-571-22055-7

Faber & Faber

The Universe Next Door. Marcus Chown—ISBN-13: 978-0747234968
Headline Paperbacks

Parallel Worlds. Michio Kaku—ISBN 0-141-01463-6
Penguin Science

About the universe being driven by recursive computer code

References:

A New Kind of Science—Stephen Wolfram *http://wolfram.com/*

About Jesus

Jesus. A.N. Wilson—ISBN 0-00-637738-6
Flamingo

About Chaos Theory—The butterfly effect.

Refer online to: http://www.imho.com/grae/chaos/chaos.html

About fossil fuel depletion

Duncan's 1996 paper on the Olduvai Theory: http://www.dieoff.org/page125.htm and his follow up paper of November 2000 http://www.dieoff.org/page224.htm

A similar paper by Jay Hanson, Energy Synops is at http://www.dieoff.org/synopsis.htm

About the power of gold and the nature of wealth

The Power of Gold, The History of an Obsession. Peter L. Bernstein—ISBN-13: 978-0470091005
John Wiley & Sons; III edition (November 5, 2004)

About death

What Happens When We Die? Dr Sam Parnia—ISBN 1-4019-0556-0
Hay House, UK, 2005,

About living inside a computer

The Matrix. DVD Warner. ASIN: B00000K19E

The Thirteenth Floor. DVD 1999. ASIN: B00004TYXU

About general science theories

Theories For Everything. ISBN 0-7922-3912-1
National Geographic

About supporting the idea of a spiritual and divine god

The Dawkins Delusion. Alister McGrath—ISBN 0-281-05927-6
SPCK—Society for Promoting Christian Knowledge

About the delusion of religion and the existence of a divine god

The God Delusion. Richard Dawkins—ISBN 0-593-05548-9
Bantam Press

About Quantum Theory—(well presented).

Introducing Quantum Theory. J.P. McEvoy & Oscar Zarate—ISBN 1-84046-057-1
Bantam Press

About becoming cyborgs

Tomorrow's People. Susan Greenfield—ISBN 0-141-00888-1
Penguin Science

About belief and proof of big ideas and questions

What We Believe But Cannot Prove. Edited by John Brockman—ISBN 10: 1-4165-2261-1
Pocket Books

Big Questions In Science. Edited by Harriet Swain—ISBN 0 09 942892 X
Vintage

About real history as it was witnessed

The Mammoth Book Of How It Happened Edited by Jon E. Lewis—

ISBN 1-84529-265-0
Robinson

A small note for the intelligent silicon-human hybrids who read this book in the future.

At first it may seem like you live in an improved world, one where virtuality and reality blur into a combined experience, which can in one moment, be personal, and the next—shared with all others. But nothing changes. Some of you will be pleasure seekers looking for the next best buzz; some will be dreamers looking for a different existence; and a few of you will be searching for the meaning behind your existence.

Before you, came billions of non-hybrids: ordinary, inferior, and now out-dated models of living creatures containing awareness. These were the best the universe could produce blindly in its desperate search to save itself. I am one of these. I am aware of the terrible struggles my forebears had back to the dawn of time, the suffering in their lives, their small joys compared to their agony of loss, pain, disease and heartache. If not for their brave struggle, I would not have been here to write this to you, and you would not be there now reading it.

On behalf of all my dead kind, I wish you well. I also ask that you, too, remember those before you, and that you carve out a good future towards a noble purpose, one which befits, and respects everything they, your forebears, endured to help you come into existence. If you live up to your true purpose, I will meet with you at the end of days, as you will meet with everyone before you. In that moment, we are all one—no longer separated by time or space; and then we can each come to know who gave so much for our common journey and goal.

Mol
2007

About the Author

Mol Smith is an artist, writer, and ex-computer programmer. He has worked in the field of Telecommunications for British Telecom, initially as a communications technician and later as a manager. He has one child, Kelly-Anne and three Grandchildren: Megan, Alice, and Maddie. He is now divorced. He currently works on a variety of projects in Oxford, and has a home in London. He has interests in Digital Art and Visual Communications, Microscopy, Biology, Philosophy, New Technologies, Video-making, 3D technologies, and History. He co-founded one of the first non-profit-making major science sites on the Internet with less budget than most people spend at their local supermarket in a month. This presence receives contributions from over 150 science authors to an online monthly ISSN registered magazine called Micscape. The web site is visited by over 100,000 people monthly and is a main teaching aid to Biology students throughout the world.

Your feedback and comment regarding his work here in this book can be made via an online forum at **www.2x21.com** and the author will reply to your own comments and questions via this channel only. You are welcomed by the author to add to the forum debates, provided your comments are constructive (either positive or negative), and not emotional rants.

Mol Smith thinks human Social Traits are the keys to humanity's success thus far, and believes that many other animals, including non-mammals, display behaviour indicating we are not the only species, sufficiently evolved, to experience emotional joy and pain.

Lesley Evans.
(Mol's companion and loved one).

"Man, unlike any other thing organic or inorganic in the universe, grows beyond his work, walks up the stairs of his concepts, emerges ahead of his accomplishments."

John Steinbeck 1902-68

www.2x21.com

Printed in Great Britain
by Amazon